T0331172

Advanced Fractal Graph Theory and Applications

This book explores the dynamic interplay between fractals and graph theory, two powerful mathematical tools with vast applications. It presents a strategic combination and the synergistic use of these disciplines to address real-world problems and challenges. This book begins with an introduction to the basic concepts of fractals and graph theory and then explores their applications in various domains, including natural phenomena modeling, scheduling, and network optimization.

This book:

- Illustrates the innovative ways in which fractals and graph theory can be combined, laying the groundwork for future applications across various industries
- Introduces the fundamental concepts and principles of both fractals and graph theory in detail, making the material accessible to a broad audience, including those new to these topics
- Explores practical applications in image processing, network optimization, social network analysis, and more, demonstrating the real-world impact of these mathematical tools
- Analyzes advanced techniques in graph theory, such as matching, domination, and coloring, with practical examples and case studies
- Highlights the latest research advancements in fractal graph theory, showcasing its potential for future developments and applications

This book is intended for students, researchers, and professionals in mathematics, computer science, engineering, and related fields.

Advanced Fractal Graph Theory and Applications

P. Tharaniya, G. Jayalalitha, Pethuru Raj,
and B. Sundaravadivazhagan

CRC Press is an imprint of the
Taylor & Francis Group, an **informa** business
A CHAPMAN & HALL BOOK

Designed cover image: ShutterStock

First edition published 2025
by CRC Press
2385 NW Executive Center Drive, Suite 320, Boca Raton FL 33431

and by CRC Press
4 Park Square, Milton Park, Abingdon, Oxon, OX14 4RN

CRC Press is an imprint of Taylor & Francis Group, LLC

ISBN: 9781032768106 (hbk)
ISBN: 9781032770697 (pbk)
ISBN: 9781003481096 (ebk)

DOI: 10.1201/9781003481096

Typeset in Times
by codeMantra

Contents

Preface

A fractal is a never-ending pattern. Fractals are infinitely complex patterns exhibiting self-similarity across different scales. They possess the innate ability to model complex physical processes and dynamic systems. The central principle of fractals is that a simple process, when repeated infinitely, can lead to highly complex outcomes.

Most fractals operate on the principle of a feedback loop, which is a process where the output of a system is fed back as input, influencing subsequent outputs. A simple operation is performed on a piece of data, which is then fed back into the system. This process is repeated multiple times, and the limit of this process is called the fractal. Fractals are predominantly self-similar, meaning that a part of the fractal is identical to the entire fractal itself.

The fractal dimension is a measure of a fractal object's complexity. It is a ratio that provides a statistical index of complexity, comparing how the detail in a fractal pattern changes with the scale at which it is measured.

Despite their complex and intriguing nature, fractals are surprisingly simple to create. They originate from a fundamental process and gradually become more intricate. Chaos theory also reflects this property, where simple processes can generate complex results. With the aid of high-performing computers, it is now possible to generate and decode fractals, presenting them in graphical representations that are easy to comprehend.

Amazingly, fractals are extremely simple to create and are found throughout nature. These repeating patterns range from the tiny branching of blood vessels and neurons to the branching of trees, lightning bolts, and river networks. Other examples include coastlines, mountains, clouds, seashells, and hurricanes. Abstract fractals, such as the Mandelbrot Set, can be generated by a computer that repeatedly calculates a simple equation. In this book, we begin with graph theory and fractal graph theory. In the third chapter, we will discuss fractal geometry, providing all the relevant theoretical and practical information to enrich our esteemed readers. The fourth chapter covers iterated function systems (IFSs), while the fifth chapter explains how to create fractals from IFSs. Fractals have a dazzling array of industrial applications, including data compression, image processing, computer graphics, and even the design of antennas and microchips, as illustrated in the sixth chapter. The seventh chapter addresses matching and its real-world applications, while the eighth chapter demystifies domination and its practical uses. The ninth chapter vividly describes the aspect of coloring and its corresponding applications. Healthcare is a promising domain for smartly applying advancements in fractal theory. The 11th chapter details how fractals are useful in circuit theory. The 12th chapter is dedicated to expounding on the advantages of fractals in architecture. Finally, the last chapter discusses fractal neural networks and their unique industrial use cases.

Authors

Dr. P. Tharaniya

In 2022, Dr. P. Tharaniya obtained her Ph.D. in the Department of Mathematics from VELS Institute of Science, Technology, and Advanced Studies, Chennai. Currently, she is working as an Assistant Professor in the Department of Mathematics at Rajalakshmi Institute of Technology, Chennai. She has over 14 years of teaching experience and more than 5 years of research experience. She has presented at more than ten international and national conferences and has attended numerous webinars, faculty development programs, and workshops related to graph theory. She has published 12 technical papers in journals and conferences. Her research interests include fractal graph theory, matching, domination, and coloring.

Dr. G. Jayalalitha

In 2011, Dr. G. Jayalalitha obtained her Ph.D. in Mathematics from Sri Chandrasekharendra Saraswathi Vishwa Mahavidyalaya. She is a lifetime member of the Indian Mathematical Society, MISTE, and the Association of Mathematics Teachers of India. Currently, she is a Professor in the Department of Mathematics at VELS Institute of Science, Technology, and Advanced Studies, Chennai. She has nearly 18 years of teaching and research experience. She has published nearly 120 articles and supervised 11 Ph.D. scholars. She has presented at more than ten international and national conferences and has served as a resource person and keynote speaker at over ten programs. Additionally, she is an advisory board member at VISTAS. Her papers have been cited nearly 158 times, and her h-index is 7. Her research interests include fractals, mathematical modeling, operations research, and graph theory.

Pethuru Raj, PhD
Vice President and Chief Architect
Edge AI Division, Reliance Jio Platforms Ltd.
Bangalore, India – 560 103.

Currently, Dr. Pethuru Raj works at Reliance Jio Platforms Ltd. (JPL), Bangalore. Previously, he worked at IBM Global Cloud Center of Excellence, Wipro Consulting Services, and Robert Bosch Corporate Research. He has over 22 years of experience in the IT industry and 8 years of research experience. He completed his CSIR-sponsored Ph.D. at Anna University, Chennai, and continued with UGC-sponsored postdoctoral research in the Department of Computer Science and Automation at the Indian Institute of Science, Bangalore. Subsequently, he was awarded two international research fellowships (JSPS and JST) to work as a research scientist for 3.5 years at two leading Japanese universities. He is a professional member of ACM and IEEE. His focus includes various digital transformation technologies such as the Internet of Things, artificial intelligence (AI), streaming data analytics, blockchain, digital twins, cloud-native computing, edge and serverless computing, reliability engineering, microservices architecture, event-driven architecture, and 5G/6G.

B. Sundaravadivazhagan, Ph.D., SMIEEE
Department of Information Technology
University of Technology and Applied Sciences-Al Mussanah, Oman

B. Sundaravadivazhagan obtained his Ph.D. in computer science from Anna University, Chennai, India. Currently, he is a faculty member in the Department of Information Technology at the University of Technology and Applied Sciences-Al Mussanah in Oman. He holds professional memberships in ISACA and ISTE and is a senior member of IEEE. His academic and research background spans more than 23 years at various institutions. He is currently working on two funded research projects for the Ministry of Higher Education Research Innovation in Oman through the TRC. His research interests include cybersecurity, AI, machine learning, deep learning, and cloud computing. He has published more than 75 technical articles in journals and conferences around the world.

1 Graph Theory – An Overview

1.1 INTRODUCTION

Mathematics plays a crucial role in various fields, such as the natural sciences, engineering, health, finance, and social sciences. The study of applied mathematics has led to the development of completely new fields of mathematics, including statistics and game theory. An arrangement that is fundamentally a collection of objects, some of which are "related" to one another, is called a graph. Graphs are commonly represented diagrammatically by a set of lines or curves that connect the edges and a set of dots or circles that represent the vertices. Graphs are one of the subjects taught in discrete mathematics. Among the key subjects covered in discrete mathematics is graph. It is established that directed graphs in which edges connect two vertices asymmetrically and undirected graphs in which edges connect two vertices symmetrically are different from one other. Graph theory, a branch of mathematics is dedicated to the study of graphs [1].

Graph theory is based on combination logic, and "graphics" are solely used to visualize data. Graph-theoretic models and applications typically involve definition and computational techniques provided by combinatorial mathematics and linear-algebra on the one hand, and linkages to the "real world" on the other hand (sometimes described in vivid graphical terms). Graph theory is fascinating to many because of this interaction [2]. Simple algorithms for planarity testing and graph drawing are presented in a section of graph theory that focuses on the graphical representation and drawing of graphs. However, this topic is addressed in a rather cursory manner; a deeper analysis would necessitate conclusions from more complex area such as curve theory and topology. This section also provides a succinct overview of matroids, a useful generalization that can be used in place of graphs.

Graph-theoretic findings and methods are typically not proven in a strictly combinatorial form; instead, they leverage the visualization opportunities provided by graphical presentations [3]. Depending on their structure, graphs can be characterized by a variety of attributes that they possess. These characteristics are defined in language peculiar to the field of graph theory. It is the number of edges in the shortest path connecting vertices U and V. Graph theory is applied across many engineering domains. For example, circuit connection design heavily relies on graph theory principles.

Topologies refer to the categories or configurations of connections. Star, bridge, series, and parallel topologies are a few types of topologies. The interactions between interconnected computers are governed by graph theory. Graphs are used to illustrate chemical and molecular structures of substances, DNA structures in living

DOI: 10.1201/9781003481096-1

FIGURE 1.1 Example of graph.

things, and even grammar and language parsing trees. They can also illustrate routes between cities. A particular type of graph known as a tree can be used to represent hierarchically organized data, such as family trees.

1.2 DEFINITIONS

1.2.1 Graph

Graphs are mathematical structures made up of a set of nodes (also known as vertices) and a set of edges. They model pair-wise relationships between items in a given collection. In a plane, edges are represented as line segments connecting the vertices, which are depicted as points.

1.2.2 Directed and Undirected Graph

A graph with undirected edges is referred to as an undirected graph. A graph with directed edges is called a directed graph.

Figure 1.2 shows the example of a directed graph.

1.2.3 Connected Graph

A graph is considered to be linked if a path connects all its vertices. In a directed graph, if directed edge is transformed into an undirected edge, the directed graph is said to have weak connectivity. If at least one vertex in a simple directed graph can be adjacent with another vertex, the graph is said to be unilaterally connected. If both vertices in a directed graph are reachable from each other, the graph is said to be highly linked.

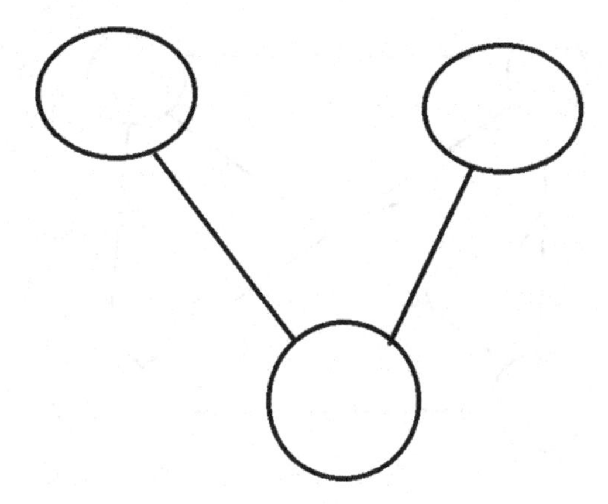

FIGURE 1.2 Directed graph and undirected graph.

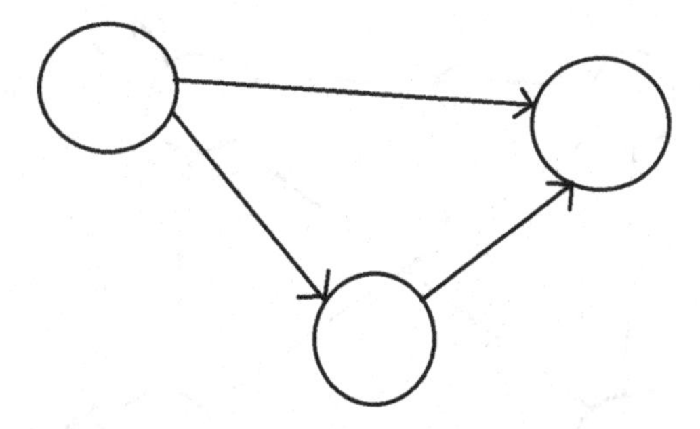

FIGURE 1.3 Weakly connected.

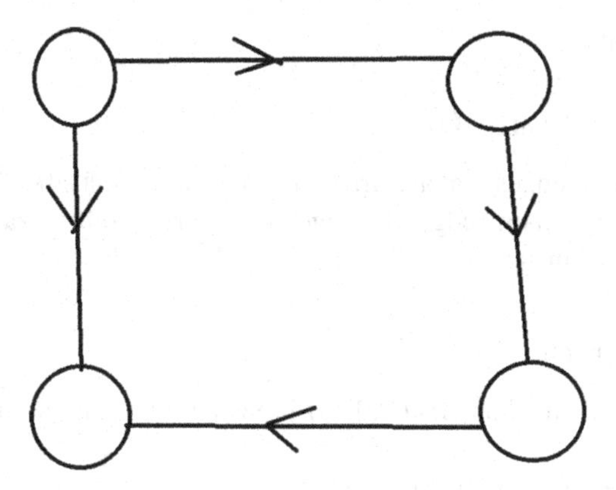

FIGURE 1.4 Unilaterally connected.

Figure 1.3 shows an example of a weakly connected graph, Figure 1.4 shows an example of a unilaterally connected, and Figure 1.5 shows an example of a strongly connected graph.

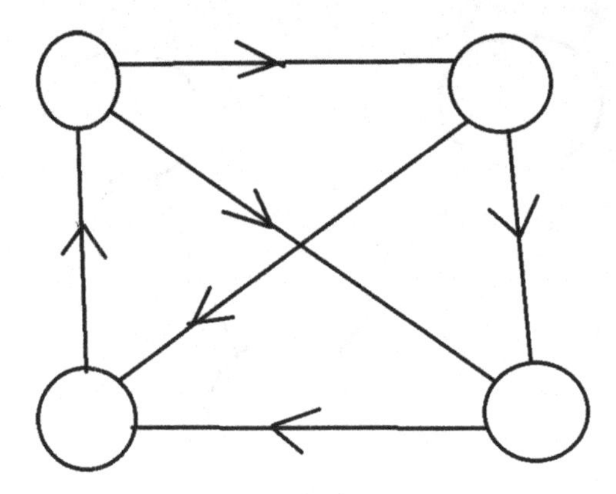

FIGURE 1.5 Strongly connected.

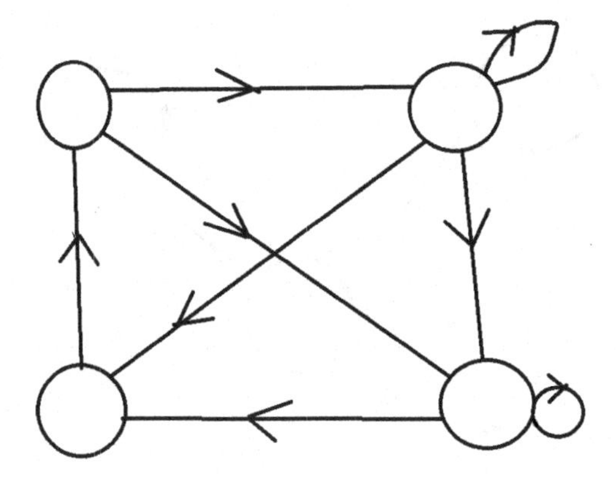

FIGURE 1.6 Loop.

1.2.4 Loop and Parallel Edges

A loop is made up of edges that are drawn from a vertex to itself. When two vertices are joined by more than one edge, the edges are referred to be parallel edges.

Figure 1.6 shows an example of a loop.

1.2.5 Simple Graph

A simple graph is defined as G= (V, E) if it has no loops and no multiple edges, or parallel edges.

Figure 1.7 shows an example of a simple graph.

1.3 EDGES AND VERTICES

A vertex, also referred to as a node (plural: vertices), is the fundamental building block of graphs. In an undirected graph, the structure consists of sets of vertices and edges, unordered pairs of vertices. In contrast, a directed graph is composed of

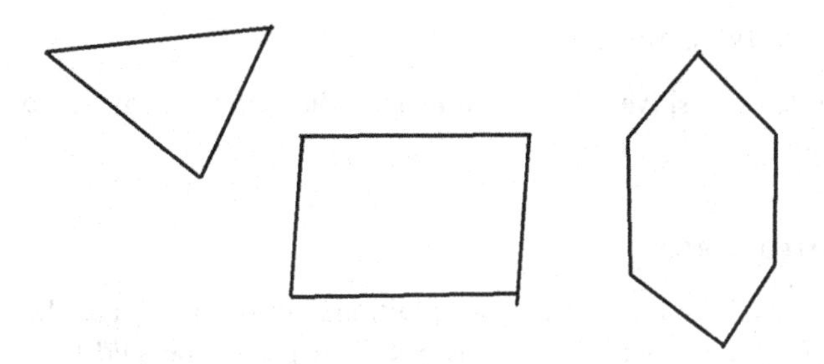

FIGURE 1.7 Simple graph.

ordered pairs of vertices called arcs. In graph diagrams, a line or arrow connecting two vertex points is commonly used to represent an edge. A vertex is usually represented as a circle with a label. The number of edges incident to a vertex is called the degree of the vertex. The number of outgoing edges is considered as the out-degree and the numbers of incoming edges is considered as the in-degree.

1.3.1 Degree of the Vertex

The degree of a vertex represented by the symbol (v) in a graph, is the number of edges incident to a vertex. In a directed graph, the out-degree, or the number of outgoing edges, is represented by $\delta+(v)$, while the in-degree, or the number of incoming edges, is represented by $\delta-(v)$.

1.3.2 Types of Vertices

A vertex with degree zero is called as isolated vertex. A vertex with one degree is called as a leaf vertex. A vertex with in-degree zero is called a source vertex, while a vertex with out-degree zero is called as sink vertex. When less than k vertices are removed from a graph, the remaining graph remains linked, which is known as a k-vertex-connected graph. If removing a vertex with all incident edges results in a subgraph with more linked elements is known as cut point. A cut edge, often called a bridge, is an edge that removed from a graph creates a new graph with more connected components. A cut set S satisfies the condition that S is a subset of E. A linked graph G becomes disconnected when its edges are removed. No proper subset of G satisfies this requirement.

1.3.3 Vertex Cover

An independent set is a set of vertices in which no two vertices are adjacent. A vertex cover is a set of vertices that contains at least one endpoint of each edge in the graph.

1.3.4 Vertex Space

The vertex space of a graph is a vector space having a set of basis vectors corresponding to the graph's vertices.

1.3.5 Vertex Transitive

A graph is vertex-transitive if it has symmetries that can map any vertex to any other vertex.

1.3.6 Labeled Vertex

A labeled vertex is a vertex that has additional information attached, making it distinguishable from other labeled vertices. Two graphs be said to be isomorphic only when the vertices between two graphs are matched based on equivalent labels. Unlabeled vertices are those that can be used in place of any other vertex in the network based on their adjacencies, without any additional information.

1.3.7 Edge Connectivity

In a connected graph, the minimum number of edges that must be removed is known as edge connectivity. $\lambda(G)$ represents the edge connectivity of a connected graph G if G is a disconnected graph, linked graph G with a bridge has an edge connectivity of 1.

1.3.8 Vertex Connectivity

In a connected graph, the minimum number of vertices that must be removed is known as vertex connectivity. The vertex connectivity of a linked graph is represented by either V(G) or k(G).

i. $E(G)=0$ if G is a disconnected graph.
ii. A linked graph G with a bridge has an edge connectivity of 1.
iii. Deleting a single vertex does not disconnect the entire graph k_n, but deleting $n-1$ vertices reduces it to a simple graph. Hence, $n-1=k(k_n)$.
iv. A graph of order at least has one vertex connectivity if and only if it has a cut vertex.
v. The vertex connectivity of a path is one

1.4 TYPES OF EDGES

Numerous data types can be represented by networks. In biological networks, the nodes represent various entities (such as proteins or genes), while the edges provide information about the connections between these nodes. We will focus on the edges first. The type of edge information determines the type of analyses that can be carried out. Therefore it is helpful to identify the main type of edges that exist in networks. Undirected edges: Typically, directed edges are represented as arrows pointing from the origin vertex—also known as the tail of the arrow toward the destination vertex – also known as the head of the arrow. Because directed graphs do not impose the restrictive requirement of symmetry in the relationships described by the edges, they are considered the most general type of graph.

Figure 1.8 illustrates an example of directed edges.

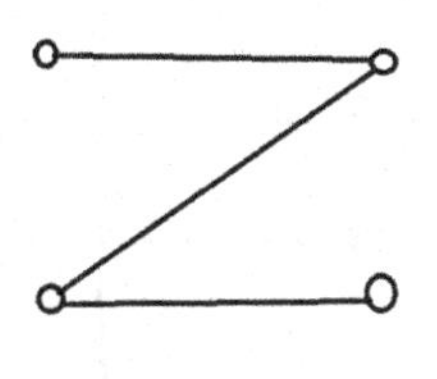

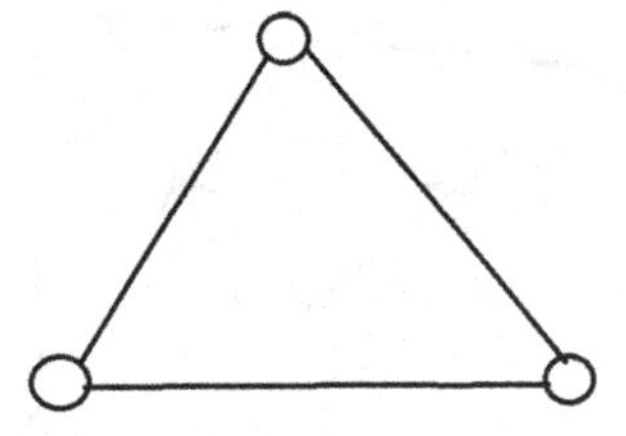

FIGURE 1.8 Undirected edges.

1.4.1 Directed Edges

In sparse networks directed graphs provide more information than similar undirected graphs. This implies that we likely to lose information if we treat a sparse directed graph as undirected. One relevant example is the construction of genealogical trees, where the relationship is "a child of" is significant. Undirected graphs are not inherently transitive, but they work well for relationships where the existence of connections is important. For instance, we can represent pedestrian pathways as an undirected graph if they allow travel in both directions.

1.4.2 Undirected Edges

Undirected edges transmit mutual understanding and connection in both directions, much like whispers between friends. They represent the soft dance of reciprocity, where power or hierarchy hold no sway over the free exchange of ideas. In a network, undirected edges embody the essence of reciprocal relationships, wherein communication is two-way and every node holds equal importance. They promote cooperation and harmony by enabling information, energy, or influence to flow effortlessly between connected nodes akin to an equal-opportunity dialogue. Undirected edges encourage the exploration of interconnectedness, nodes engage in both giving and receiving, resulting in an interdependent web that strengthens the network's structure. In the creative domain, undirected edges ignite the flow of inspiration and ideas, launching cooperative endeavours that transcend individual boundaries. As nodes interact and influence one another in a symphony of shared invention, they invite investigation and discovery. Ultimately, undirected edges serve as a reminder of the beauty of reciprocity and mutual respect, where relationships flourish in an environment equality and openness, allowing communication to flow freely. They subtle highlight the beauty of connection and the strength of teamwork in shaping the fabric of our shared experiences.

Figure 1.9 will provide an example of undirected edges.

1.4.3 Weighted Edges

Numerical values can be assigned to both directed and undirected edges. This is used to illustrate concepts such as gene-on-gene interacting, the quantitative difference in expression that one gene causes over another, or the degree of sequence similarity between two genes. Additionally weights based on edge centrality values and various other topological factors can be applied

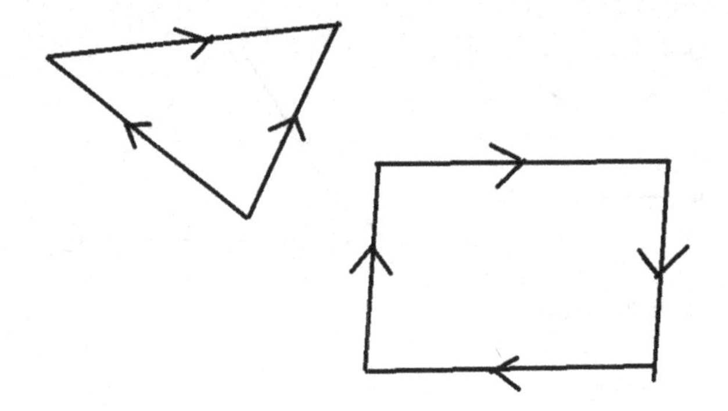

FIGURE 1.9 Directed edges.

1.5 FUZZY GRAPH

Graphs often do not accurately represent all systems due to uncertainty or ambiguity in the parameters of those systems. Crisp graphs and fuzzy graphs are structurally similar. However, fuzzy graphs are particularly important when there is uncertainty regarding vertices and / or edges. A fuzzy graph $\xi = (V, \sigma, \mu)$ is an algebraic structure consisting of a non-empty set V together with a pair of functions $\sigma : V \rightarrow [0,1]$ and μ: $V \times V \rightarrow [0,1]$ such that for all x, y $\in$ V, $\mu(x, y) \leq \sigma(x) \wedge \sigma(y)$ and μ is a symmetric fuzzy relation on σ. Here $\sigma(x)$ and $\mu(x, y)$ represent the membership values of the vertex x and of the edge (x, y) in ξ respectively. The fuzzy graph $\xi_1 = (V, \sigma_1, \mu_1)$ is called a fuzzy sub graph of $\xi = (V, \sigma, \mu)$ if $\sigma_1(x) \leq \sigma(x)$ for all x and $\mu_1(x, y) \leq \mu(x, y)$ for all edges (x, y), x, y $\in$ V.

1.5.1 Strongest Fuzzy Graph

For the fuzzy graph $\xi = (V, \sigma, \mu)$, an edge (x, y), where x, y $\in$ V is called strong if $1/2\ [\{\sigma(x) \wedge \sigma(y)\}] \leq \mu(x, y)$ and it is called weak otherwise. The strength of an edge (u, v) is denoted by $I\ (u, v) = \mu\ (u, v) / \sigma\ (u) \wedge \sigma\ (v)$.

The strength of a path is defined as min $\{\mu\ (x_{i1}, x_i), i=1,2,3.....,n\}$. In other words, the strength of a path is the weight (membership value) of the weakest arc of the path. The strength of connectedness between two nodes x and y is defined as the maximum of the strengths of all paths between x and y.

A function F defined on some set X with real or complex values is called bounded, if the set of its value is bounded [12]. In other words, there exists a real number M such that $|f(x)| \leq M$ for all x in X. If the sequence a_n is either monotone increasing or monotone decreasing, then a_n is said to be monotone. If the sequence a_n is monotone increasing and bounded above, then a_n converges. Likewise, if an is monotone decreasing and bounded below then a_n converges. This thesis explains the existence of fuzzy fractals in many fields. Fuzzy number, which is an extension of real numbers, has properties that can be related to the theory of numbers. It is widely used in engineering applications because of its suitability for representing uncertain information

1.6 HAMILTON GRAPH

A graph with a Hamiltonian cycle is known as a Hamiltonian graph. A Hamiltonian cycle visits each vertex in the graph exactly once, with the beginning and ending vertices being the same. In other words, a Hamiltonian cycle is a closed loop that circles each graph vertex precisely once. There is no straightforward prerequisite for a graph to be Hamiltonian. However, several well-known types of Hamiltonian graphs exist, such cycle graphs and complete graphs, which are graphs in which every pair of distinct vertices is connected by an edge. A path in a graph that visits each vertex exactly once is called a Hamiltonian path.

A graph is referred to as traceable if it has a Hamiltonian path but not necessarily a Hamiltonian cycle. Applications for Hamiltonian graphs can be found in many domains, including optimization, network architecture, and computer science. For instance, in the traveling salesman problem, finding the shortest path that makes exactly one stop in each city before returning to the starting point corresponds to identifying a Hamiltonian cycle in a weighted graph. As fundamental concepts in graph theory and combinatorial optimization, Hamiltonian graphs and cycles have a wide range of theoretical and practical applications. Despite the challenges in determining whether a graph is Hamiltonian, the study of Hamiltonian graphs remains a hot topic in computer science and mathematics research.

1.7 THE ORIGIN OF THE GRAPH THEORY

Graph theory delves into the examination of connections among objects, depicted through vertices and interconnecting lines known as edges. Such structures are commonly referred to as graphs. Originating in the 18th century, the evolution of graph theory can be traced through various mathematical problems and puzzles. A concise historical overview is provided below. The 18th-century quandary emerged in the city of Konigsberg (present-day Kaliningrad, Russia). The city, spanning both banks of the Pregel River, featured two sizable islands linked by seven bridges. The puzzle posed the question of whether one could stroll through the city, traversing each of the seven bridges precisely once, and ultimately returning to the initial point. Geography of Konigsberg: The city comprised four distinct areas, two riverbanks and two islands, linked by a total of seven bridges. The identified land masses were Kneiphof, Lomse, and the two riverbanks.

Bridge A: Connecting the two riverbanks
Bridge B: Connecting one of the riverbanks to Kneiphof.
Bridge C: Connecting Kneiphof to the other riverbank
Bridge D: Connecting Kneiphof to Lomse
Bridge E: Connecting Lomse to one of the riverbanks
Bridge F: Connecting Lomse to the other riverbank

The task involved discovering a path within the city that would traverse each bridge precisely once and lead back to the initial point. This challenge captivated the residents of Konigsberg and transformed into a widely embraced puzzle within the city.

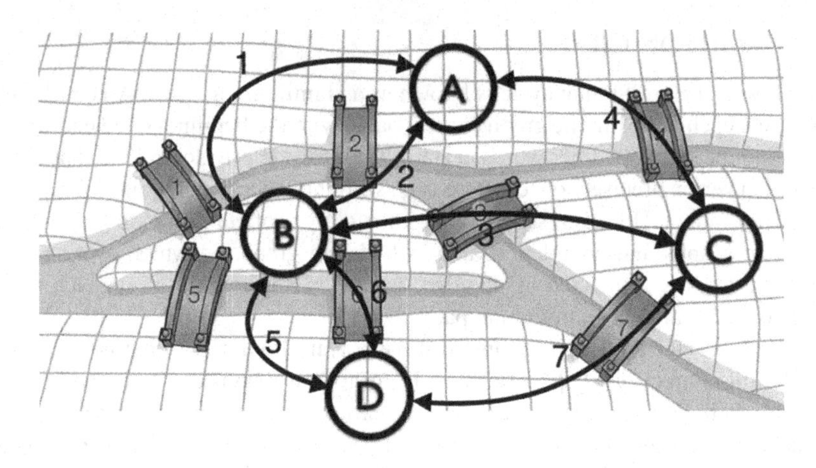

FIGURE 1.10 Konigsberg bridge.

Euler's Seven Bridges of Konigsberg (1736): Graph theory is commonly credited to Leonhard Euler, a Swiss mathematician, who in 1736 successfully tackled the renowned Seven Bridges of Konigsberg problem. Euler approached the challenge uniquely by transforming the city's layout into a graph. His abstraction proved beyond dispute that it is impossible to find a path across the city that crosses each bridge exactly once.

This ground-breaking solution served as the cornerstone for graph theory, and Euler's paper on the matter, titled "Solutio problematis ad geometriam situs pertinentis" (The Solution of a Problem Relating to the Geometry of Position), published in 1736, is widely regarded as the birth of the field. Euler's pioneering work introduced the fundamental concepts of graph theory, establishing it as a distinct and significant branch within mathematics. In 1736, Leonhard Euler published a significant paper titled that the solution of a Problem Relating to the Geometry of Position. This publication holds a pivotal place in the history of mathematics, as it established the groundwork for graph theory. This field, stemming from Euler's foundational work, has since been extensively applied across various disciplines. Background: The issue Euler tackled in his paper stemmed from a widely known puzzle in Konigsberg (present-day Kaliningrad, Russia). The puzzle revolved around the possibility of strolling through the city, crossing all bridges precisely once, and returning to the initial point. Euler approached this challenge by innovatively shifting the perspective, abstracting the city's physical layout into a mathematical structure.

Figure 1.10 illustrates the structure of the Konigsberg Bridge.

1.7.1 Euler's Approach

Euler represented the land masses (islands and riverbanks) as vertices and the bridges as edges. This abstraction led to the creation of a mathematical object that we now call a graph. **Vertices and Edges**: Euler assigned symbols to each land mass and bridge. The four land masses were represented by vertices labeled A, B, C, and D while the bridges were represented by edges. A represents one of the riverbanks, B represents Kneiphof (an island), C represents the other riverbank, and D represents Lomse (the second island).

Edges (Bridges):

1. AB: Bridge connecting the first riverbank (A) to Kneiphof (B)
2. AC: Bridge connecting Kneiphof (B) to the other riverbank (C)
3. BC: Bridge connecting Kneiphof (B) to Lomse (D)
4. CD: Bridge connecting Lomse (D) to the other riverbank (C)
5. DB: Bridge connecting Lomse (D) to the first riverbank (A)
6. BA: Bridge connecting Kneiphof (B) back to the first riverbank (A).
7. CA: Bridge connecting the other riverbank (C) back to the first riverbank (A)

Euler subsequently expressed the problem using the graph and investigated its characteristics, placing particular emphasis on Eulerian paths and circuits. This abstract portrayal enabled Euler to extend his solution, establishing the foundation for the emergence of graph theory. The specific graph used for the Königsberg scenario became a notable illustration in history, and Euler's observations regarding connectivity and paths within graphs significantly influenced the field of mathematics.

1.7.2 Network Analysis

Euler examined the connections between the vertices and edges, transforming the tangible problem into a concern about the structure of this conceptual graph. Eulerian Paths and Circuits: Euler introduced the notions of an Eulerian path and an Eulerian circuit. An Eulerian circuit is a closed path that includes each edge exactly once. Impossibility of the Desired Walk: Euler demonstrated that the challenge of discovering a route through Königsberg. Euler's publication marked the inception of graph theory as an independent field in mathematics. By introducing abstract structures and expressing the problem in terms of vertices and edges, Euler laid the foundation for a robust mathematical framework that extended beyond the confines of the Königsberg problem. This conceptualization empowered mathematicians to construct a comprehensive theory applicable to diverse problems concerning relationships and connectivity. Legacy: The Königsberg Bridge problem and Euler's resolution laid the groundwork for the establishment of graph theory as a formal discipline. Today, graph theory is an essential component of discrete mathematics, finding practical applications in computer science, network analysis, operations research, and various other fields. Euler's contributions are revered as a pivotal force in the history of mathematics, playing a foundational role in shaping contemporary graph theory. Kirochhoff's Matrix Tree Theorem:Gustav Kirchhoff employed matrices to depict graphs in his exploration of electrical networks. His Matrix Tree Theorem, introduced in 1847, offered a powerful method for enumerating spanning trees.

Figure 1.11 shows an example of the Konigsberg Bridge.

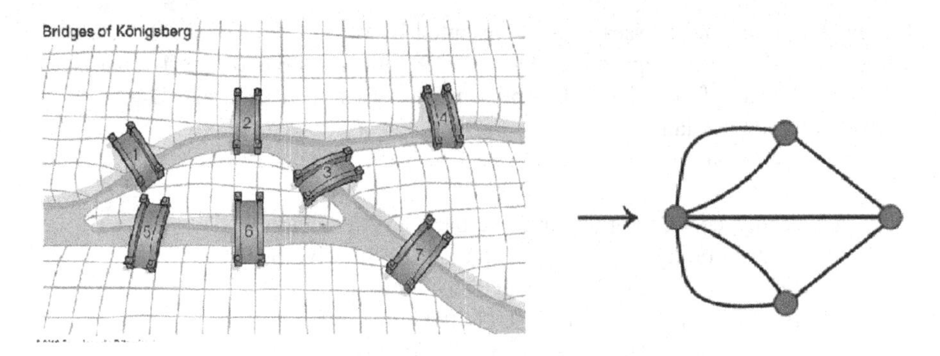

FIGURE 1.11 Euler solution to the Konigsberg problem.

1.8 GRAPH THEORY THROUGH THE CENTURIES

1.8.1 Graph Theory in 19th century

During the 19th century, mathematicians like August Möbius and Listing made significant contributions to topology, which encompasses the study of spaces and their properties. Graphs, particularly those representing surfaces, played a major role in enhancing the understanding of the properties associated with these spaces.

1.8.1.1 Networks and Transportation Planning

While Euler's contributions marked the inception of graph theory, the 19th century saw further progress and utilization of these principles. It is noteworthy that the term "graph theory" may not have been widely used during this era. Advancements in transportation, such as the establishment of railways and telecommunication networks, were particularly significant. Informal applications of graph theory concepts were evident in the planning and optimization of these networks. Engineers and planner, tasked with designing and organizing efficient routes and connections for transportation systems, likely employed graph theory principles, even if they were not explicitly identified as such.

1.8.1.2 Map coloring

The concept of map coloring, which is connected to graph theory, found practical applications in political geography. In the 19th century, the four-color theorem, a result from graph theory, answered the question of how to color every map using only four colors, ensuring that adjacent regions are colored distinctly. This theorem is significant for the practical aspects of creating maps and flags, as it allows for the design using a minimal color palette.

1.8.1.3 Electrical Network Analysis

In the 19th century, physicist Gustav Kirchhoff, contributed to the advancement of electrical circuit theory. Although Kirchhoff's laws are primarily associated with circuit analysis, his research explored the use of matrices to depict networks of electrical elements. The principles he employed connections to later concepts in graph theory, particularly in the analysis of network structures.

1.8.1.4 Combinatorics

While the specific term "graph theory" may not have been widely used in the 19th century, the foundational ideas of graph theory were actively explored and applied across various disciplines. The formalization and recognition of these ideas occurred in the 20th century, driven by the field's expansion and the development of specific terminology and notation

1.8.2 Graph Theory as a Formal Field (20th Century)

Although the specific term "graph theory" was not coined until this era, mathematicians in the 19th century explored into several mathematical concepts that are now associated with graph theory. During this period, mathematicians such as Frank Harary and Paul Erdős made notable contributions to the formalization and development of graph theory as a distinct field.

1.8.2.1 Map Coloring and the Four Color Theorem:

During the mid-19th century, mathematicians and cartographers engaged with the challenge of map coloring. The Four-Color Theorem, initially proposed by Francis Guthrie in 1852, emerged as a significant result in graph theory. This theorem posits that any map on a plane can be colored using only four colors, ensuring that adjacent regions do not share the same color. Although the proof of the theorem did not come until the 20th century, the problem and its investigation involved the examination of planar graphs and coloring.

1.8.2.2 Hamiltonian Paths and Circuits

Hamiltonian paths involve traversing each vertex exactly once, while Hamiltonian circuits are closed paths that include every vertex exactly once. Although Hamilton's primary focus was not explicitly on graphs, his ideas served as foundational elements for subsequent developments in graph theory. Mathematicians continued to investigate these concepts to solve problems related to the traversability of graphs and networks.

1.8.2.3 Networks and Mathematical Physics

In the 19th century, advancements were made in applying mathematics to the physical sciences. Although not explicitly utilizing graph theory, mathematicians like Gustav Kirchhoff employed principles of network theory in the context of electrical circuits, later establishing connections to graph theory. It is crucial to highlight the systematic development and formalization of graph theory. Mathematicians in the 19th century often delved into problems and concepts that laid the groundwork for subsequent developments in graph theory. The terminology and notation associated with graph theory were established in the decades that followed.

1.9 APPLICATIONS AND GROWTH OF GRAPH THEORY

With the increasing prevalence of computers, graph theory has found applications in computer science, operations research, network analysis, and various other fields. Researchers have developed algorithms and methods to address real-world problems using graph theory concepts. Today, graph theory constitutes a fundamental aspect of discrete mathematics with widespread applications across various disciplines, including computer science, social network analysis, biology, and logistics. The field continues to evolve through ongoing research, with new applications emerging regularly.

1.9.1 Graph Theory in Image Processing

In image processing, graph theory is frequently employed to depict and examine the connections among pixels or regions within an image. Utilizing graph-based representations offers a robust framework for capturing the structure of an image and extracting significant information.

The following provides a concise overview of the application of graph theory in image processing.

Figure 1.12 shows an example of a grayscale image and a binary image.

1.9.1.1 Graph Representation

The type of connectivity (e.g. four-connectivity or eight-connectivity) depends on the application.

1.9.1.2 Segmentation

Graph theory applied in image segmentation, aiming to divide an image into significant regions or objects. In the context of segmentation algorithms, the image is often represented as a graph, with each segment treated as a connected component within the graph.

1.9.1.3 Image Representation

Representing images as graphs involves assigning pixels to vertices and indicating relationships between neighboring pixels with edges. This representation facilitates the extraction of structural information from the image.

FIGURE 1.12 Grayscale image and binary image.

1.9.1.4 Graph Cuts

Utilizing graph cut algorithms, such as the min-cut/max-flow algorithm, is common in image segmentation. The image is graphically represented, and cuts in the graph correspond to segmenting the image into distinct regions. Min-cut algorithms assist in finding the optimal partitioning of the graph.

1.9.1.5 Image Denoising

Graph-based methods play a role in image denoising by treating noisy pixels as outliers in the graph. Employing graph-based filtering techniques allows for the effective identification and suppression of noisy pixels.

1.9.1.6 Graph-Based Filters

The application of graph filters to an image helps in smoothing or enhancing specific features. For instance, bilateral filtering can be implemented using graph structures to preserve edges while reducing noise.

1.9.1.7 Object Recognition

Graph theory aids in object representation and recognition within images. Graph-based models capture pixel relationships, enabling algorithms to identify patterns or shapes based on graph connectivity.

1.9.1.8 Graph Based Image Retrieval

Graph-based representations allow for indexing and retrieval of images based on similarity measures between graphs. This enables the retrieval of images with similar structures or patterns. In summary, graph theory serves as a versatile framework for various image processing tasks, enabling the modeling of complex relationships, segmentation, denoising, and the extraction of meaningful information.

Figure 1.13 shows an example of the process of filtered graph.

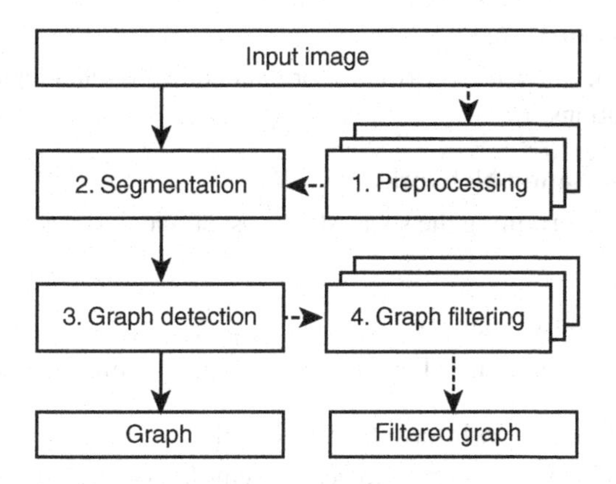

FIGURE 1.13 Segmentation, image representation, graph cuts, and graph based image retrieval.

1.9.2 Application of Network with Graph Theory

Graph theory plays a vital role in analyzing and optimizing diverse networks, with applications across different fields.

1.9.2.1 Social Networks

Community Detection: Identifying groups or communities within social networks based on connections between individuals.

Centrality Measures: Evaluating the importance of nodes within a social network using centrality measures.

1.9.2.2 Computer Networks Routing Algorithms

Designing and analyzing routing algorithms for optimal data transmission paths in computer networks.

1.9.2.3 Network Topology Design

It is Optimizing node layout and connectivity for efficiency and reliability in computer networks.

1.9.2.4 Financial Networks

Analyzing financial transactions and identifying patterns in transaction networks.

1.9.2.5 Wireless Networks

Designing and optimizing communication in networks without a fixed infrastructure.

1.9.2.6 Wireless Sensor Networks

Analyzing connectivity and coverage of sensor networks in various applications

1.9.2.7 Epidemiology Models

Modeling disease spread in populations using graph structures.

1.9.2.8 Internet of Things

It is designing efficient sensor networks for data collection and analysis in Internet of Things applications.

1.9.2.9 Transportation Networks

Route Planning: Determining the shortest or most efficient routes in road, rail, or air networks

1.9.2.10 Traffic Flow Analysis

Studying and optimizing traffic flow in road networks to minimize congestion.

1.9.2.11 Protein–Protein Interaction Networks

Analyzing protein interactions in biological systems to understand cellular processes.

1.9.2.12 Telecommunication Networks

Designing and analyzing the performance of telecommunication networks using graph theory.

1.9.3 Graph Theory with AI

Graph theory expresses intricate linkages and structures, making it a vital tool in many branches of artificial intelligence (AI). In AI, graph theory can be utilized in the following ways.

1.9.3.1 Data Representation

Various data structures, including social networks, biological networks, recommendation systems, and knowledge graphs, can be represented using graphs. Edges signify the connections between nodes, which represent entities. Understanding these relationships is essential for tasks such as social network analysis and recommendation systems in AI applications.

1.9.3.2 Algorithms for Searching

In AI, graph traversal algorithms such as depth-first search (DFS) and breadth-first search (BFS) are essential for navigating extensive state spaces. Pathfinding tasks, including route planning, solving puzzles, and making decisions in games, utilize these methods.

1.9.3.3 Optimization Issues

Optimization issues, such as minimal spanning tree and shortest path finding, employ graph techniques like Dijkstra's algorithm. AI leverages these algorithms for logistics planning, resource allocation, and route optimization.

Clustering and Classification: In AI applications like image segmentation, document clustering, and recommendation systems, related data points are grouped using graph-based clustering algorithms such as spectral clustering and community discovery techniques.

1.9.3.4 Deep Learning on Graphs

Graph neural networks (GNNs), which extend conventional neural networks, operate directly on data organized into graphs. In AI, GNNs have proven effective for tasks including graph classification, link prediction, and node classification, particularly in fields like recommendation systems, social network analysis, and drug discovery. Knowledge Representation and Reasoning: Knowledge graphs organize information using nodes and edges. In AI, knowledge graphs facilitate tasks such as inference, question answering, and semantic search. Algorithms for graph-based reasoning assist in drawing logical conclusions and generating new information from existing knowledge graphs. Fraud Detection and Anomaly Detection: Graph-based anomaly detection algorithms identify unusual behaviors or patterns in network traffic or financial transactions. These algorithms support cybersecurity and fraud detection by analyzing the graph structure and spotting abnormalities.

1.9.3.5 Natural Language Processing

Natural language processing (NLP) employs dependency graphs and semantic graphs to depict the syntactic and semantic relationships between words in a phrase. Graph-based NLP models use these representations for tasks such as semantic parsing, part-of-speech tagging, and named entity recognition. By utilizing graph theory,

AI systems can model intricate relationships, make defensible judgments, and extract insightful knowledge from interconnected data, thereby enabling the development of more intelligent and practical solutions across various fields.

1.9.4 Graph Theory in Agriculture

Graph theory can be applied in agriculture in several ways to increase productivity, manage resources, and optimize operations. Here are a few examples:

1.9.4.1 Crop Planning and Rotation

Fields and crops can be represented as nodes and edges, respectively, in graphs. Graph algorithms can optimize crop rotation schedules to promote soil health, avoid pest infestations, and maximize yields over time by analyzing the relationships and compatibility between various crops.

1.9.4.2 Irrigation Network Optimization

Graphs with nodes representing fields or irrigation points and edges representing water pipes or channels can model agricultural irrigation systems. Graph algorithms can efficiently distribute water to different crops based on their water requirements, optimize the irrigation network structure, and reduce water waste.

1.9.4.3 Supply Chain Management

Production facilities, distribution hubs, retail stores, and transportation routes can be represented as nodes in graphs, which simulate supply chains in agriculture. By analyzing the supply chain graph, agricultural enterprises can ensure timely delivery of produce to markets, minimize transportation costs, and improve transportation routes.

1.9.4.4 Management of Pests and Diseases

The spread of pests and diseases in agricultural fields can be represented graphically. Graph algorithms can forecast the growth of infestations and suggest targeted interventions, such as pesticide treatments or quarantine measures, to limit outbreaks by examining the connectivity between various fields and the movement patterns of pests or pathogens.

1.9.4.5 Farm Management System

Graph databases are useful for storing and displaying intricate interactions among various components of a farm management system, including fields, crops, workers, equipment, and environmental conditions. Through graph database queries, farmers can gain insights into crop performance, profitability, and resource use, facilitating data-driven decision-making and improving farm operations

Overall, graph theory offers powerful tools and methods for deciphering complex agricultural systems, streamlining processes, and enhancing the sustainability and productivity of farming practices.

1.9.5 Application of Graph Theory in Industries

Graph theory can represent intricate linkages and structures, making it applicable across various industries. Its applications in the industry are as follows:

1.9.5.1 Network Analysis

In sectors such as logistics, transportation, and telecommunications, graphs are used to simulate different types of networks. Graph algorithms enable capacity planning, routing, and optimization of network infrastructure. For instance, in telecommunications, graph theory enhances bandwidth efficiency, reduces signal interference, and optimizes communication network topology.

1.9.5.2 Management of Supply Chains

Supply chains can be visualized as graphs, where nodes represent various production and distribution phases, and edges signify the movement of products or information between them. Graph algorithms ensure timely product delivery to clients, reduce transportation costs, and optimize supply chain logistics.

1.9.5.3 Social Network Analysis

Graph theory is frequently applied in social network analysis to examine the connections between people or entities across various sectors, including marketing, finance, and healthcare. In social networks, graph algorithms assist in identifying communities, recognizing key nodes, and predicting user behavior [4].

1.9.5.4 Recommendation Systems

In recommendation systems, user-item interactions are modeled using graphs. Nodes represent users and items, while edges depict interactions or connections between them. Graph-based recommendation algorithms analyze relationships between users and items to generate personalized recommendations [6].

1.9.5.5 Fraud Detection

Graph theory is used in fraud detection to analyze relationships between entities, such as customers, transactions, and accounts. In sectors like banking, insurance, and e-commerce, graph algorithms help identify suspicious patterns, prevent financial losses, and detect fraudulent activities.

1.9.5.6 Drug Development

Chemical compounds and their interactions can be visualized as graphs, with nodes representing atoms or molecules and edges representing chemical bonds or interactions. Graph-based algorithms optimize drug candidates for safety and efficacy, aiding drug discovery and development by studying chemical structures and predicting drug-target interactions [5].

1.9.5.7 Cybersecurity

Computer and device networks can be represented as graphs, with nodes signifying individual devices and edges representing communication links or routes. Cybersecurity applications utilize graph-based algorithms to detect malicious activity, identify anomalies, and protect against cyber threats such as malware, phishing, and network attacks.

1.9.5.8 Semantic Web and Knowledge Graphs

In Semantic Web and knowledge graph technologies, data and information are represented as graphs of connected entities and relationships, based on graph theory.

Graph-based knowledge representation facilitates advanced search, data integration, and semantic reasoning in sectors like publishing, e-commerce, and healthcare.

In conclusion, graph theory provides powerful methods and tools for simulating, assessing, and refining intricate systems and networks across diverse sectors, promoting increased productivity, creativity, and decision-making.

1.9.6 Graph Theory in Ornaments

Graph theory has multiple applications in decoration design and analysis.

1.9.6.1 Pattern Generation

Graph theory is a useful tool for creating complex ornamental patterns and designs. In a graph, edges signify connections or links between nodes, which represent individual elements or motifs. By employing techniques such as recursive subdivision or random walks, designers can produce ornament patterns that are both visually appealing and distinctive.

1.9.6.2 Symmetry Analysis

Many decorations rely heavily on symmetry. Graph theory provides tools for examining the symmetry characteristics of ornamental patterns. Graphs that have nodes for motif locations and edges for symmetry transformations can illustrate symmetrical patterns. By analyzing the symmetries inherent in the graph structure, designers can comprehend and control the symmetry characteristics of ornaments.

1.9.6.3 Tessellation and Tiling

The tessellations and tilings of geometric shapes serve as the foundation for many ornaments. The fundamental structure of tessellations can be represented using graph theory, where nodes stand for tiles and edges represent adjacency interactions between them. Techniques such as graph coloring and minimal spanning trees enable designers to create tessellations with the uniformity and visual appeal they desire.

1.9.6.4 Fractal Ornaments

Self-similar patterns with intricate detail at various scales are known as fractals. Fractal ornament patterns can be modeled and analyzed using graph theory. Graphs with nodes representing elements at different scales and edges showing transformations of self-similarity can illustrate fractal patterns. By utilizing graph-based fractal algorithms such as recursive subdivision or iterative refinement, designers can produce visually striking fractal ornaments.

1.9.6.5 Analysis of Complexity

Graph theory offers methods for evaluating the complexity of an ornament design. An ornament's complexity can be assessed using graph-based metrics like graph entropy, degree distribution, or density. By examining these metrics, designers can evaluate ornament designs for visual richness, intricate detail, and aesthetic appeal

Overall, graph theory provides valuable methods and insights for the analysis and design of ornaments, empowering designers to create aesthetically pleasing and structurally sound decorative patterns.

1.9.7 Chemical Graph Theory

Within the field of mathematical chemistry, chemical graph theory applies concepts from graph theory to the study of molecular structure. In this field, molecules are represented as graphs, with chemical bonds depicted as edges (connecting lines between nodes) and atoms as vertices (nodes). The goal of chemical graph theory is to use these graph representations of molecules to investigate and understand various features of molecules.

This involves examining molecular graphs for patterns and themes that may indicate specific chemical properties or behaviors, assessing molecular connectivity, determining molecular symmetry, predicting molecular stability, and analyzing molecular reactions. The application of graph theory in chemistry facilitates the development of computational techniques and algorithms to tackle complex chemical problems, such as designing new compounds, predicting their properties, and understanding their interactions. It provides a robust mathematical framework for comprehending the composition and behavior of molecules.

1.9.8 Graph Theory with Chemical Components

Graph theory can be applied in various ways to evaluate and understand chemical components. Figure 1.14 illustrates the chemical formula of an alkane molecule.

1.9.8.1 Molecular Structure Analysis

In accordance with graph theory, atoms can be represented as vertices and bonds as edges in a graph. Important structural characteristics, such as ring systems, branching patterns, and overall symmetry, can be deduced by examining the topology and atomic connectivity of the molecular graph.

1.9.8.2 Isomer Enumeration

Isomers are compounds that have distinct structural configurations but share the same chemical formula. Graph theory can be employed to generate all non-isomorphic graphs corresponding to valid molecular structures, thereby listing all possible isomers of a given chemical formula.

1.9.8.3 Substructure Looking

This method involves searching for specific molecular fragments or patterns within larger chemical structures using graph theory. Substructure searching is essential in chemical informatics, drug discovery, and chemical synthesis planning.

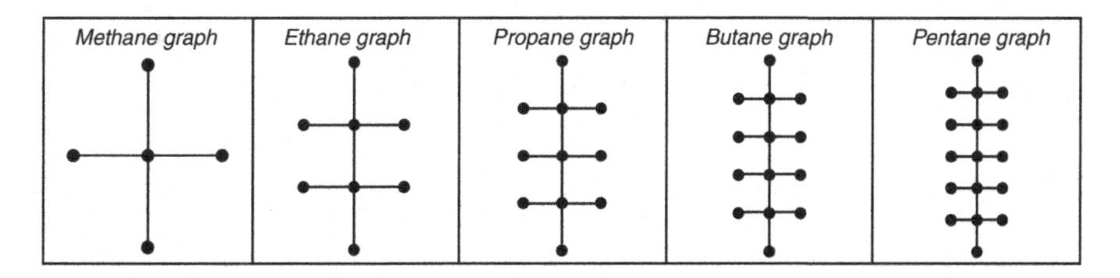

FIGURE 1.14 The structural formula of an alkane molecule with topological index.

1.9.8.4 Molecular Descriptors

These numerical representations, known as graph-based molecular descriptors, are derived from the molecular graph. They represent various topological and structural characteristics of molecules and are utilized in virtual screening, molecular modeling, and studies of quantitative structure–activity relationships (SARs).

1.9.8.5 Networks of Chemical Reactions

Molecular graph transformations can be used to illustrate chemical reactions. By visualizing the network of chemical reactions as a graph, researchers can analyze reaction pathways, identify key intermediates, and understand the kinetics of chemical processes.

1.9.8.6 Chemical Similarity and Clustering

The similarities between molecules are calculated based on their graph representations using graph-based techniques. Similarity metrics derived from graph theory are employed in scaffold hopping in drug development, molecular similarity searching, and grouping molecules based on structural similarity. Overall, graph theory provides a flexible framework for dissecting the structural, topological, and dynamic features of chemical constituents, thereby shedding light on their characteristics, interactions, and behaviors.

1.9.9 Carbon Structures with Graph Theory

Since carbon is a key element in organic chemistry, graph theory is often utilized to explore carbon structures. The following is an application of graph theory to carbon structures: An alkane graph is a tree where the edges represent carbon-carbon or hydrogen-carbon bonds in an alkane, and the vertices represent atoms. An alkane is defined as an acyclic saturated hydrocarbon, which is a molecule made up of carbon and hydrogen atoms arranged in a tree structure with only one carbon-carbon bond. Figure 1.15 shows an example of carbon structures.

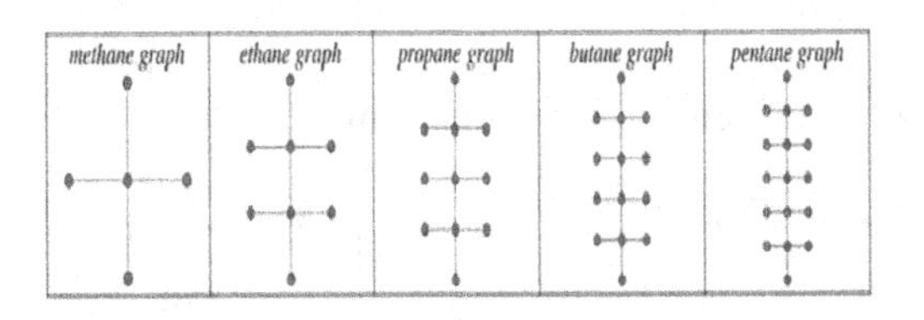

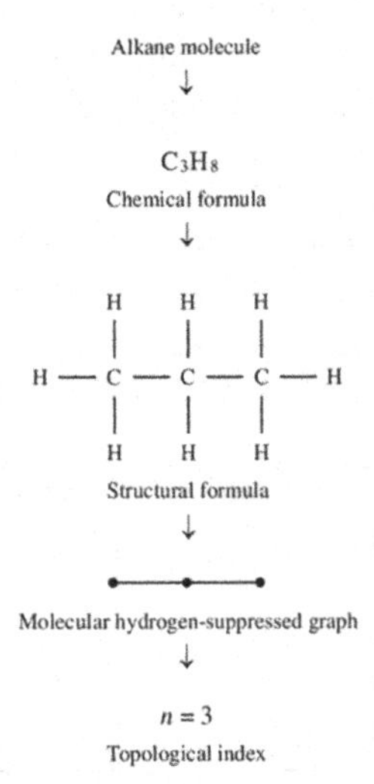

FIGURE 1.15 Example of carbon structures.

1.9.9.1 Representation as Graphs

Chemical bonds (such as single, double, or triple bonds) between carbon atoms are the edges of graphs that depict carbon structures, such as organic compounds. This format enables simplified viewing and study of intricate molecular structures.

1.9.9.2 Topological Analysis

Carbon structures can be analyzed topologically using graph theory. This entails examining the relationships between carbon atoms, recognizing rings and cycles in the molecular graph, and describing the molecule's overall architecture.

1.9.9.3 Isomer Enumeration

Graph theory can be applied to list and categorize isomers of carbon compounds. Isomers are compounds that have distinct structural configurations but the same chemical formula. A given molecular formula can be used to generate and analyze all potential non-isomorphic graphs, which can help detect and classify various types of isomers, including constitutional isomers, stereoisomers, and tautomers.

1.9.9.4 Aromaticity and Conjugation

Graph theory can be used to study aromatic carbon structures, such as benzene and its derivatives. Analyzing the connectivity and resonance structures of the molecular graph can provide insights into aromaticity, which is associated with the presence of conjugated pi electron systems. The stability and reactivity of aromatic rings can be understood and identified with the aid of graph-based techniques.

1.9.9.5 Substructure Searching

Substructure searching in carbon structures utilizes graph theory. Graph matching algorithms are employed to look for specific structural motifs or patterns within larger chemical networks. These patterns include functional groups, substituents, and reaction sites, and are represented as subgraphs. Substructure searches are crucial in chemical informatics, drug discovery, and SAR research.

1.9.9.6 Molecular Descriptors

Graph-based descriptors measure the structural and topological characteristics of carbon compounds. These descriptors are valuable for predicting the physical, chemical, and biological properties of organic molecules, as they capture details about molecular size, shape, branching, and symmetry. In conclusion, graph theory provides a robust framework for examining the composition, characteristics, and reactivity of carbon compounds, enhancing our understanding of the intricate and diverse field of organic chemistry.

1.9.10 Graph Theory with Arts

There are interesting uses for graph theory in many artistic fields. This is how it relates to the arts: Graph theory is frequently employed by artists to create network art, where nodes and edges represent objects and their relationships, respectively.

These networks can depict conceptual links as well as social interactions and communication patterns. Network art explores themes of interconnectivity, intricacy, and emergence and can take various forms, such as interactive installations, visualizations, and installations.

1.9.10.1 Algorithmic Art

Algorithmic art applies mathematical concepts to produce artistic forms and patterns, drawing inspiration from graph-based algorithms. Creative use of algorithms for traversing graphs, such as BFS or DFS, can lead to the creation of complex visual compositions. Artists can investigate the aesthetic properties of graph topologies—such as symmetry, balance, and rhythm—to produce visually striking works of art.

1.9.10.2 Generative Art

Ideas from graph theory often inspire generative art, which employs algorithms to autonomously create artwork. Artists can utilize graph-based procedural generation techniques to create a diverse range of shapes, textures, and patterns. For example, graph coloring algorithms can generate intricate color patterns, while graph grammars can be employed to produce organic shapes or architectural structures.

1.9.10.3 Data Visualization

Artists commonly use data visualization techniques to explore and convey abstract themes, with graph theory being fundamental to the presentation of large datasets. By utilizing graph layout methods and representing data as graphs, artists can create visually engaging representations that communicate information in an intuitive and aesthetically pleasing manner. Data-driven artworks frequently address contemporary themes like social networks, urban dynamics, and environmental trends

1.9.10.4 Interactive Installations

Interactive artworks often incorporate graph theory concepts to create captivating and immersive experiences. Artists can develop interactive exhibits that allow users to manipulate graph topologies through digital or physical interfaces. These installations invite viewers to actively engage with the artwork and influence its evolution while exploring ideas such as collaboration, communication, and emergence.

1.9.10.5 Mathematical Art

Graph theory serves as a rich source of inspiration for artists who work with graph structures to investigate their visual complexity and beauty. Graph-theoretic concepts such as fractals, tessellations, and graph embedding can inspire artists as they create intricate drawings, sculptures, or digital artworks. Mathematical art highlights the intrinsic beauty of mathematical structures, celebrating the interaction of geometry, topology, and aesthetics. In conclusion, graph theory provides a rich environment for artistic inquiry, fostering a variety of artistic expressions that combine mathematical precision with aesthetic awareness. By bridging the gap between art and science, graph theory enhances our understanding of the expressive potential of artistic practice and the structural complexity of networks. Figure 1.16 shows an example of art created using graph theory.

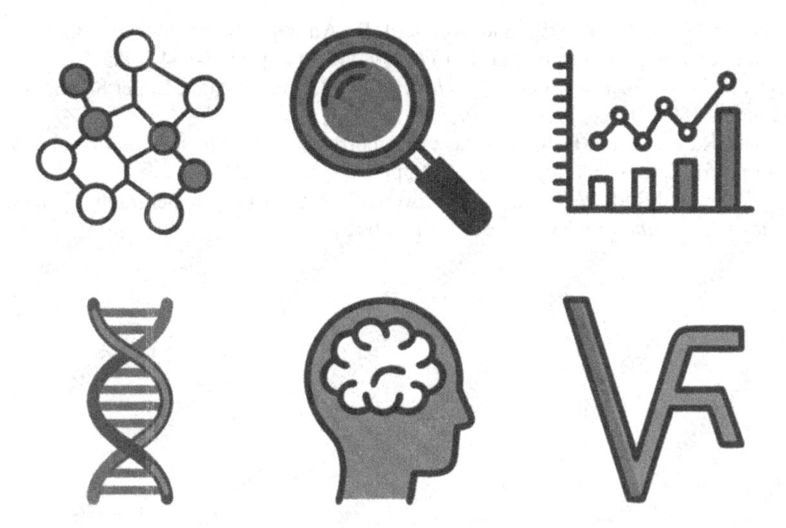

FIGURE 1.16 Example of art with graph theory.

1.10 CONCLUSION

Graph theory provides a rich tapestry of concepts and tools that are applicable to many different domains, making it a cornerstone of both computer science and mathematics. It offers a unifying framework that transcends disciplinary boundaries, serving a variety of purposes - from modeling complex systems to optimizing transportation routes, deciphering the intricacies of chemical structures, and understanding social networks. Fundamentally, graph theory embodies the beauty of abstraction; it simplifies intricate systems into sophisticated diagrams of vertices and edges. Its practical relevance and theoretical depth empower researchers, engineers, and innovators to tackle real-world challenges with accuracy and insight, contributing to its versatility. As we explore the complex networks of interconnection that shape our environment, graph theory continues to illuminate the path toward new discoveries and creative solutions. Its influence extends across academia, business, and other domains, shaping our understanding of networks, interactions, and systems.

Ultimately, graph theory stands a testament to the beauty of interconnectivity, the strength of abstraction, and the limitless possibilities of human inquiry. It is much more than just a branch of mathematics. It is a constant companion that guides us toward greater comprehension and boundless opportunities as we navigate new territories and forge new connections.

REFERENCES

1. Bondy, J. A., and Murty, U. S. R., *Graph Theory with Applications*, North-Holland, New York (1976).
2. Bresar, B., On Vizing's conjecture, discussiones mathematicae, *Graph Theory*, Vol. 21(1), pp. 5–11 (2001).

3. Cockayne, E. J., Gamble, B., and Shepherd, B., An upper bound for the k-domination number of a graph, *Journal of Graph Theory*, Vol. 9(4), pp 533–534 (1985).
4. West, D. B., *Introduction to Graph Theory*, Vol. 2, Prentice Hall, Upper Saddle River, NJ, pp. 1–150 (2001).
5. Kumar, R., and Kant, R., Graph theory, importance and scope, *International Global Journal for Research Analysis*, Vol. 4(7), pp. 365–367 (2015).
6. Frankl, P., and Rödl, V., Near perfect coverings in graphs and hypergraphs, *European Journal of Combinatorics*, Vol. 6(4), 317–326 (1985).

2 Fractal Graph Theory

2.1 INTRODUCTION

People are excited by beautiful self-similar images known as Fractals [1]. Long before the invention of computers, individuals were studying fractals. The concept of fractals originated when people attempted to measure the length and boundary of the British Coast. They used large-scale maps, which represented only half the length of the coastline measured on proper maps. Consequently, mathematicians introduced fractals as powerful tools for measuring complex phenomena, solving problems in computer science, and uncovering secrets from a wide variety of systems. Figure 2.1 shows that an example of Fractal Graph.

2.1.1 History of Fractals

Fractals have improved our ability to evaluate and classify random or organic objects, whether they are proper or not. Some mathematicians still believe that a function of randomness exists that cannot be perfectly described by mathematical equations. In 1872, at the Royal Prussian Academy of Sciences, Karl Weierstrass presented the first definition of a function along with a graph that had the unusual property of being everywhere continuous but nowhere differentiable. Today, this

FIGURE 2.1 Examples of fractal graph.

DOI: 10.1201/9781003481096-2

is considered a fundamental aspect of fractals. In 1883, Georg Cantor explained Weierstrass's function which is a subsets of the real line known as the Cantor set, and highlighted its unusual properties. Later, he expanded on the ideas on the ideas of Poincare and expressed dissatisfaction with Weierstrass's abstract and analytic definition. He provided many geometric definitions and derivations, particularly through hand-drawn images featuring self similar functions, one of which is known as the Koch Snowflake. In 1915, Waclaw Sierpinski created a famous triangle with similar analytic definition and self-similarity property, known as Sierpinski Triangle or Gasket. Pierre Fatou and Gaston, Julia were also working on similar concepts and describing many results simultaneously [2].

2.1.2 Father of Fractal Geometry

French-American mathematician Benoit B. Mandelbrot (1924–2010) of Polish descent is best known for developing the Mandelbrot set and for his groundbreaking work in fractal geometry. He and his family fled the onset of Nazism in 1936, have been born in Warsaw, Poland. Mandelbrot developed the idea of fractals in the 1960s, driven by his fascination with the irregularities and structures found in nature. Fractals are intricate geometric forms that exhibit self-similarity at various scales, or at any level of magnification. The field of fractal geometry, which has applications in computer graphics, biology, physics, mathematics, and art, was founded on Mandelbrot's research [3].

The Mandelbrot set is known for its complex border, which, when magnified, reveals an endless amount of information. It has been utilized in science, technology, and art, becoming a symbol of popular culture. Mandelbrot's research transformed our understanding of asymmetrical forms and patterns observed in the natural world, significantly influencing a wide range of disciplines, including computer graphics and chaos theory. Among the many books and research articles authored was "The Fractal Geometry of Nature," which popularized the concept of fractals. Mandelbort's legacy continues to inspire scholars, artists, and enthusiasts alike. In 1975, Mandelfort coined term 'Fractal' which had taken from the latin word 'Fractus'. It is meant that broken or fractured. He used his mathematical definitions in conjunction with computer based construction and visualization.

2.2 FRACTAL DIMENSION

Fractal Dimension is utilized in Machine Learning as a method of dimensionality reduction [4]. It helps illustrate how scaling affects a model or object. Consider a very complex shape; that is graphed to scale and reduced. This type of work can be measured and evaluated using fractal dimension. Mathematicians have often been misled by nature; while nature is imaginative, 19th-century mathematicians were not. – As cited by Benoît Mandelbröt, F. J. Dyson, *The Divinatory Character of Geometry.* The final and most remarkable characteristic of fractals is that they exist as intermediate three, two, and one-dimensional objects. Fractional dimensions appear align naturally with the pattern found in nature, and we should embrace this concept as well. To begin, let's define what we mean when we say "dimension". Mathematical definitions of "dimension" vary widely but maintain a consistent core. A shape's dimension can be intuitively understood as a score that indicates how well the shape occupies its surrounding space or as a measure of its roughness.

We can achieve mathematical precision for these intuitive concepts. Consider a piece of paper as an example of a fractional dimension. It is (practically) two-dimensional. A solid sphere occupies more space than the paper, as it is three-dimensional. Now, if we crumple the paper into a ball, we create a shape that resembles a fractal, which consists more area than the flat paper, but less than the solid sphere. This crumpled receives a dimension score of about 2.5.

In comparison, your lungs are roughly 2.97 dimensional, due to their fractal geometry, which allows them to fit a small volume (equivalent to a few tennis balls) within a large surface area (comparable to a few tennis courts). This extensive surface area enables you inhale enough oxygen to sustain your life, Fractals are present in everything around you, from the structure of the universe on the largest scales to the delicate form a fern. Your brain is one of the fractal elements of your anatomy. Once you become aware of fractals, you'll be surprised by how many different settings they appear in within everyday life including clouds, vegetation, landscapes, church windows, and laboratories. In addition to helping us model natural forms, fractal mathematics can rekindle our sense of awe for the world we live in.

2.2.1 Complexity from Simplicity

The British mathematician Michael Barnsley, now affiliated with the Australian National University, examined fractals from a distinct perspective while Mandelbröt was studying them. One notable characteristic of fractals is that, despite their infinitely complex geometry, their complexity arises from very simple basic fundamental definitions. A small set of mathematical mappings accurately describes how the smaller copies are assembled to produce the entire fractal shape. An application called the Chaos Game, featured in Barnsley's seminal 1988 book Fractals Everywhere, allowed computers to generate any fractal shape from its known mappings very quickly [5]. The Chaos Game tracked the movement of point as it jumped around from a starting position in space, with each hop determined by a random selection from among the mappings. It was remarkable to observe how quickly the point would converge onto a "strange attractor" – the fractal shape – regardless of its starting point or which mappings it traversed. Once there, it would oscillate around the attractor indefinitely.

The core of Chaos Theory lies in these fractal attractors. The infinite intricacy of fractal shapes means that even the smallest nudge to the system can displace the point entirely off the attractor, as the behaviour of a chaotic system also oscillates around a fractal attractor. Figure 2.2 shows the fractal structures of fern leaves

Most importantly, Barnsley discovered a method to compute the list of fractal mappings for any specified shape. His methods played a key role in the development of image compression, allowing the first version of Microsoft Encarta to fit tens of thousands of photos onto a single CD by fully reconstructing complex structure from simpler maps.

FIGURE 2.2 Fractal structures of fern leaves.

Beautiful structures known as fractals are created by nature and can be seen all around us. Although it is difficult to characterize them precisely, most share four common characteristics infinite intricacy, zoom symmetry, complexity from simplicity, and fractional dimensions. These characteristics will be discussed in more detail below. If you stop to take a closer look at the next fern you encounter, you will find an excellent example of these traits. First, take note of the fern's finely detailed shape. It's fascinating how much the leaves resemble miniature versions of the branches. In reality, the entire fern is primarily composed of the same fundamental form replicated at progressively smaller scales. The most astounding discovery is that fractal mathematics reveals that this commonplace fern leaf is actually an intermediate two-dimensional shape rather than a one-dimensional one. We struggle to provide a straightforward response because traditional high school geometry taught us Euclidean geometry. There are very few regular shapes in the natural world, despite cylinders and rectangles being effective for modeling technological form.

Figure 2.3 shows an example of Amazing Leaves, which resemble a miniature version of the branches.

2.2.2 Infinite Intricacy

Unlike the work of Euclid, certain natural patterns are so random and fragmented that they exhibit an entirely different degree of complexity, not merely a higher one. When Beniot Mandelbrot discovered the first fractal in history in 1861, the mathematical community was taken aback. Grab a pen and create a zigzag pattern;

FIGURE 2.3 Amazing leaves: resembles a miniature version of branches.

the result should be multiple pointed corners connected by lines. To demonstrate that such a pattern could exist, the German mathematician Karl Weierstrass produced the ultimate mathematical staccato, a zigzag so jagged that it was reduced to only corners. No matter how many times the design was extended, all semblance of a smooth line would eventually disappear into an infinite grid of corners that were packed closer and closer together.

The uneven elements present in Weierstrass's shape at all sizes were the first thing that set it apart as a fractal shape. Mathematicians referred to Weierstrass's shape as "pathological" since it deviated from the well-tried established calculus tools that had been painstakingly developed over the previous few centuries. It was merely a tantalizing glimpse of a completely new form of shape until mathematicians gained access to modern computing power and the keys to the promised land. The relatively new field of mathematics known as fractals has significantly impacted on the development of many aspects of contemporary life. The scientific term for the branch of fractal geometry that investigates into and analyses the application of fractal features and geometry is fractal theory.

The first characteristic that distinguished Weierstrass's shape as a fractal was the uneven parts present at all sizes of the shape. Weierstrass's shape was labelled as "pathological" by mathematicians because it was differed from the tried and true calculus instruments that had been laboriously crafted over the previous few centuries before. Until mathematicians gained access to modern processing power and the keys to the promised Land, it was simplify a tantalizing glimpse of an entirely new form of shape, Fractals, a relatively young field of mathematics, have greatly influenced the evolution of many aspects of modern life. Fractal theory is the scientific term for the area of fractal geometry that investigates and examines the applications of fractal characteristics and geometry.

2.3 IMPORTANCE OF FRACTAL GRAPH THEORY

In graph theory, fractals are important because they provide information about the self-similarity, structure and complexity of different graph-based systems [6]. The significance of fractals in graph theory can be summarized as follows: Fractals are known for their hierarchical structure characterized by self-similarity at various sizes. This concept parallels the idea of hierarchy in graphs, where larger structures emerge from smaller-scale patterns that are repeated. Fractal graphs provide a foundation for understanding hierarchical order in complex networks, such as social networks, biological systems, and the internet

2.3.1 Scale-free Networks

Fractals are linked to scale-free networks, which are defined by highly connected nodes or hubs as indicated by a power-law degree distribution. Many real-world systems, such as the World Wide Web, citation networks, and protein interaction networks are characterized by scale-free networks. By examining the fractal features of graphs, researchers can gain insight into the evolution of scale-free behaviour and its effects on network resilience, information flow and robustness.

2.3.2 Graph Embedding

Using fractals, graphs can be embedded into spaces that maintain self-similarity. Fractal embedding's enable a compact representation of large making graph data storage, compression and visualization more efficient. Additionally, they allow for the analysis of graph features at various resolutions, which can reveal underlying patterns and structures that are not always visible in conventional graph representations.

2.3.3 Fractal Dimension

A fractal's degree of complexity or self-similarity is measured by its fractal dimension. The concept of fractal dimension is utilized in graph theory to quantify the complexity of graph topologies and to describe their topological characteristics. By examining the fractal features of graphs, researchers can better understand the evolution of scale-free behaviour and its implications for network resilience, information flow, and robustness. Fractal dimension analysis illuminates the underlying organizational principles and emergent behaviour of nodes by examining their connectedness, clustering, and geographic distribution.

2.3.4 Generate Graphs at Random

Random graphs with fractal-like characteristics are produced by fractal-based models. By simulating network growth and evolution, these models capture fundamental dynamics and generate synthetic networks characterized by hierarchical organization, scale-free degree distributions, and small-world properties. Random graph creation based on fractals is employed to simulate real-world network characteristics, bench mark algorithms and investigate how structural aspects affect network dynamics.

2.3.5 Applications in Complex systems

Fractals in graph theory are beneficial across various complex systems, including neural networks, ecological networks, and urban systems. Bu utilizing fractal principles, researchers can gain insights into the dynamics of complex systems, their resistance to disturbances, and their capacity to adapt to changing contexts. In complex systems, their resistance to disturbances, and their capacity to adapt to changing contexts. In complex systems, the fractal analysis of graph-based data can reveal phase transitions, critical events, and hidden patterns. In conclusion, fractals are integral to graph theory they provide a framework for understanding the dynamics, organization and structure of intricate networks. Their broad applicability offers significant insights into the interdependence and self-organization of systems at different scales. By incorporating fractal principles into graph theory, scholars enhance their understanding of complex phenomena and facilitate innovative solutions to practical problems.

2.3.6 Relation between Graph Theory and Fractal Graph Theory

Fractal graph theory builds on the principles of classical graph theory to analyse the features and structures of fractal graphs. The two fields are closely linked with fractal graph theory, expanding upon the ideas and methods of classical graph theory, particularly in the analysis of graphs with fractal-like properties. This relationship facilitates the examination of connectivity, topology and algorithms associated with fractal graphs.

In graph theory, fractals are important because they provide insight into the self-similarity, structure and complexity of various graph-based systems. The significance of fractals in graph theory is as follows: they are characterized by a hierarchical structure, marked by self-similarity different scales. This concept parallels the idea of hierarchy in graphs, where larger structures emerge from smaller-scale patterns that are repeated. Fractal graphs offer a foundation for understanding hierarchical organization in complex networks, such as social networks, biological systems, and the internet. Fractals are linked to scale-free networks, which feature highly connected nodes or hubs as indicated by a power-law degree distribution. Many real world systems, such as the World Wide Web, citation networks, and protein interaction networks are characterized by scale free properties. By examining the fractal features of graphs, researchers can gain insights into the evolution of scale-free behaviour and its effects on network resilience, information flow, and robustness.

2.3.6.1 Graph Embedding

Using fractals, graphs can be embedded into spaces that maintain self-similarity. Compact representations of large graphs are made possible by fractal embedding's, which also enhance graph data storage, compression and visualization. Additionally, they enable the analysis of graph features at various resolutions, uncovering underlying patterns and structures that may not be evident in conventional graph representations.

2.3.6.2 Fractal Dimension

A fractal object's degree of complexity or self-similarity is measured by its fractal dimension. The concept of fractal dimension is used in graph theory to quantify the complexity of graph topologies and describe their topological characteristics. By examining the fractal features of graphs, researchers can gain insights into the evolution of scale-free behaviour and its effects on network resilience, information flow, and robustness. Analyzing the connectedness, clustering, and geographic distribution of nodes in networks, fractal dimension analysis illuminates on the underlying organizational principles and emergent behaviours of these nodes.

2.3.6.3 Generate Graphs at Random

Fractal-based models produce random graphs with fractal-like characteristics. By simulating network growth and evolution, these models capture the fundamental dynamics and generate synthetic networks with hierarchical organization, scale-free degree distributions, and small-world characteristics. Random graph generation based on fractals is employed to simulate real-world networks, benchmark algorithms, and investigate how structural aspects influence network dynamics.

2.3.7 Applications in Complex Systems

Graph theory's fractals are useful in a variety of complex systems, including neural networks, ecological networks, and urban systems. By utilizing fractal principles, researchers can gain insights into the dynamics of these systems, their resistance to disturbances, and their ability to adapt to changing contexts. Fractal analysis of graph-based data in complex systems can reveal phase transitions, crucial events, and hidden patterns. In conclusion, fractals are essential to graph theory as they provide a framework for understanding the dynamics, organization, and structure of intricate networks. Their broad applicability offers important insights into the interdependence and self-organization of systems across different scales. By incorporating of fractal principles into graph theory, scholars enhance their understanding of complex phenomena and facilitate innovative solutions to real-world problems.

2.3.8 Relation between Graph Theory and Fractal Graph Theory

Fractal graph theory builds on the principles of classical graph theory to analyze the features and structures of fractal graphs. Graph theory and fractal graph theory are closely linked subjects. Here is how the two are related:

Basis for Graph Theory: Fractal graph theory extends the ideas and methods of classical graph theory, enabling the analysis of graphs with fractal-like properties. It relies on foundational concepts of classical graph theory, which are essential for examining the connectivity, topology, and algorithms of graphs.

2.3.9 Research on Fractal-Like Structures

The field of fractal graph theory focuses on graphs that exhibit characteristics similar to those of a fractal. The defining feature of these graphs is the recurrence of patterns at various scales, akin to the self-similarity observed in fractal geometries. Fractal graph theory investigates the topological and structural characteristics of fractal graphs, such as their hierarchical structure, scaling behaviour, and dimensionality, by extending concepts from fractal geometry to graphs.

2.3.10 Analysis of Fractal Dimension

To measure the level of self-similarity or complexity of fractal graphs, fractal graph theory integrates the concept of fractal dimension. By assessing a graph's space-filling ability, the fractal dimension reveals its underlying topological and geometric characteristics. Fractal dimension analysis serves as useful tool for understanding the structural complexity of fractal graphs and characterizing their scaling behaviour. Fractal graph theory investigates techniques for constructing fractal graphs through recursive algorithms or iterative procedures. These methods iteratively add or refine parts of the graph according to established rules, producing graphs with fractal-like features. This construction enables the creation of artificial graphs that exhibit self-similarity at various scales, facilitating the investigation of intricate network architectures.

2.3.11 Applications in Complex Systems

Fractal graph theory is instrumental in modeling and analyzing complex systems, including social, biological, and geographical networks that exhibit fractal organization. By utilizing fractal graph representations, researchers can gain insights into the hierarchical structures, connection patterns, and scaling characteristics of these complex systems. Fractal graph provides a solid foundation for understanding the emergent behaviours and dynamics of complex networks across various fields. In conclusion, fractal graph theory applies the fundamental concepts of classical graph theory to the analysis of graphs with fractal-like characteristics, offering fresh perspectives on the scaling behaviour and structural complexity of intricate networks. By incorporating ideas from fractal geometry into graph theory, fractal graph theory establishes a robust frame work for examining for dynamics and organization of complex systems at multiple scales.

2.4 FEATURES

A fractal is defined as "a rough or irregular geometric shape that can be partitioned into small pieces, each of which is a similar to reduce size copy of the whole. Generally, a fractal as a geometric object has the following features:

- It has a proper structure at various small scales
- It is too irregular to be described in traditional Euclidean geometric language.
- It has self-similarity property
- It has a Hausdorff dimension that exceeds its topological dimension
- It has a simple and recursive definition.

2.5 CLASSIFICATION OF FRACTALS

Fractals can be classified into three types based on the ratio of their self-similar properties. Self-similarity means that each part of the whole set appears the same at various scales.

FIGURE 2.4 Example of exact self-similarity.

2.5.1 Exact Self Similarity Fractals

This is the most precise type of self-similarity object. It strongly satisfies the self-similarity condition with the fractal appearing identical at various scales. This self-similar object is accurately similar to each of its parts. Scale invariance refers to the principle that allows for the measurement of length, energy, etc. It represents the ideal form of self-similarity at any magnitude, where smaller part of the object resembles the whole. Fractals defined by iterated function systems demonstrate exact self-similarity. For example, Sierpenski triangle and Koch snowflake exhibit exact self-similarity. In nature, various instances, such as fern, leaves, cauliflower, and broccoli, display this property. Each part of fern is identical to the whole fern, and each frond branch is equal to the entire frond. Cauliflower is also partitioned into smaller parts, each of which appears similar. Figure 2.4 shows an example of exact self-similarity.

2.5.2 Quasi-Self-Similarity

This is a less strict form of self-similarity. These fractal objects appear identical at different scales approximately but not exactly. Quasi-self-similar fractals contain small copies of the entire fractal in deformed and manipulated forms. They are defined by recurrence relations but do not exhibit exactly self-similarity. For example, the mandelbrot set is quasi self-similar, as the satellites are approximations of the entire set but not exact copies. Figure 2.5 shows an example of quasi-self-similarity.

FIGURE 2.5 Example of quasi self-similarity.

FIGURE 2.6 Example of statistical self-similarity.

2.5.3 Statistical Self-Similarity

This is the weakest type of self-similarity. The fractal exhibits numerical or statistical measures that are preserved across scales. Most definitions of "fractal" trivially imply some form of statistical self-similarity. Fractal dimension itself is a numerical measure that is preserved across scales. Random fractals are examples of fractals that are statistically self-similar, but neither exactly nor quasi-self-similar. The Coastline of Britain serves as another example, one cannot expect to find microscopic version of Britain by examining a small section of coast with a magnifying glass.

Figure 2.6 illustrates statistical self-similarity.

2.5.4 Random Fractals

The application of random fractals has two major criteria. First, it involves the modelling and characterization of natural process. Second it addresses the practical issue of computer simulation of such phenomena. The first application is aimed at that those interested in nature, providing accurate models that encompass all relevant aspects of the same phenomena. The goal is to enhance understanding of natural phenomena to make predictions possible. This analysis and modeling directly relate to nature, physics, chemistry, biology, etc. The aim of the second application is the direct intimation of nature. A key expectation is computational speed and efficiency coupled with realistic output. Fractional Brownian motion is the newest model which has proven most useful for characterizing shapes and processes with fractal appearances. Geological evaluation of a landscape differs from fractal approximation as it is used to describe natural processes mathematically while exposing substantial literature and traditional implementation.

2.6 GENERATING FRACTALS

Fractals are complex, self-similar structures generated through iterative procedures that repeat patters at various scales. Here's a simple overview of the process or creating fractals: Iteration function systems (IFS) are frequently used to create fractals such as the Julia set and the well-known Mandelbrot set. This technique involves taking a starting set of points and repeatedly performing a series of affine transformation on it. Each transformation has an associated probability that indicates the likelihood of its application. As the iteration proceed, the points coverage towards the fractal shape defined by the IFS. Fractals like the Burning Ship fractal and the Mandelbrot set are often produced using escape-time algorithms which repeated apply a mathematical function to every point on a complex plane. Three common techniques for generating fractals are given as follows: Figure 2.7 presents an example of random fractals.

FIGURE 2.7 Example of random fractals.

2.6.1 Escape-Time Fractals

Escape-Time Fractals are defined by a recurrence relation at each point in a space. Examples of this type include the Mandelbrot set, the Julia set, the Burning Ship fractal and the Lyapunov Fractal. These fractals are computed by repeatedly applying a transformation to a given point in the plane. The resulting series of transformed points is referred to as the orbit of the initial point. An orbit diverges when its points grow further apart without bounds allowing a fractal to be defined as the set of points whose orbit does not diverge. For instance, consider the transformation $z' = z^2$ in the complex plane C. Recursively applying this formula to a complex point z generates orbits that converge to 0 when $|z| < 1$, diverge when $|z| > 1$, and trace perfect circle when $|z| = 1$.The orbits of the fractal generated by the transformation $z' = z^2$ are relatively easy to predict. However, the situation becomes much more complex when an arbitrary constant is added to the transformation function. Specifically, consider the transformation $z' = z^2 + c$ (where c is any complex number).

Figure 2.8 illustrates an example of an escape-time fractal.

2.6.2 Iterated Function Systems

Iterated function systems are characterized by a fixed geometric replacement rule. Examples of such fractals include the Cantor set, Sierpenski carpet, Sierpenski gasket, Peano curve, Koch snowflake, Harter-Heighway dragon curve, T-square, and Menger sponge.

FIGURE 2.8 Example of escape-time fractal.

2.7 FRACTAL GENERATING SOFTWARE

Fractal-generating software encompasses any graphics software that creates images of fractals. Mobile apps are available for users to play with or explore fractals. Some programmers develop fractal software for personal use due to its novelty. The generation of fractals has posed significant challenges in pure mathematics. Fractal-generating software produces mathematical beauty through visualization and can simulate natural landscapes using fractal landscapes and scenery generation programs. This software often presents users with a limited range of settings and features, relying on pre-programmed variables; it is possible to replicate natural landscapes with fractal landscapes and scenery generation software adding regularity to an otherwise sterile computer-generated scene. Screensavers, wallpaper generators, and music visualization applications also produce fractals; this software offers constrained set of capabilities and settings, often relying on pre-programmed variables. It is possible to replicate natural landscapes with fractal landscapes and scenery generation software, adding irregularity to an otherwise sterile computer- generated scene. Screensavers, wallpaper generators, and music visualization applications also produce fractals. This software offers constrained set of capabilities and settings, often relying on pre-programmed variables. Fractals are frequently used in the demo scene because simple formulas can yield complex graphics. As generating fractals like the Mandelbrot set, is computationally intensive, these graphics are commonly used for benchmarking devices. Throughout the 1990s, the increasing adoption of computers with floating-point units in central processing units or math co-processors contributed to the popularity of creating fractal graphics. Rendering high-resolution VGA standard graphics can now take several hours.

Algorithms for fractal formation demonstrate remarkable parallelizability. Consequently, the use of graphics processing units in computers has significantly accelerated rendering times and enabled real time, parameter adjustments, which were previously hindered by render delays. Fractal-generating software has been redesigned to utilize multi-threaded processing. Online, 3D fractal generation first appeared in 2009. For the book, an early inventory of fractal-generating software was compiled.

2.8 CONCLUSION

In conclusion, the combination of graph theory and fractals provides an intriguing approach to investigating complicated structures across various domains. Fractals can be examined through the lens of graph theory to better understand their complex geometries and self-similar patterns. This analysis focuses on connectedness, adjacency, and other graph-theoretic features. This multidisciplinary method offers valuable tools for comprehending and simulating a wide range of phenomena, from technological networks to natural systems. By harnessing the rich mathematical frameworks of graph theory and fractals, researchers can discover new applications and perspectives that foster. Further investigation is necessary to fully understand the connection between fractals and graph theory, particularly in areas such as network dynamics, optimization, and algorithm design. Overall, the synthesis of graph theory and fractals holds great potential for advancing our understanding of complex systems and promoting interdisciplinary collaboration in research and engineering.

REFERENCES

1. Andres, J., and Rypka, M., Fuzzy fractals and hyperfractals, *Fuzzy Sets and Systems*, Vol. 300, pp. 40–56 (2016).
2. Barnsely, M. F., *Fractals Everywhere*, 2nd ed., Academic Press, Boston, MA (1993).
3. Berry, M. V., Lewis, Z. V., and Nye, J. F., On the Weierstrass Mandelbrot fractal function, *Proceedings of the Royal Society of London. A. Mathematical and Physical Sciences*, Vol. 370(1743), pp. 459–484 (1980).
4. Falconer, K. J., *Techniques in the Fractal Geometry*, John Wiley & Sons, Ltd, Chichester (1997).
5. Garg, A., A review on natural phenomenon of fractal geometry, *International Journal of Computer Applications*, Vol. 975, p. 8887 (2014).
6. Telcs, A., Spectra of graphs and fractal dimensions, *Journal of Theoretical Probability*, Vol. 8, pp. 77–96 (1995).

3 Fractals and Fractal Geometry

3.1 INTRODUCTION

In 1970, Beniot Mandelbrot developed fractal geometry [1]. A fractal is a subset of topological dimensions or Euclidean space characterized by a fractal dimension, which extends the concept of the Hausdorff dimension. Fractal Geometry shapes exhibit the same object at various scales. This field lies within the study of measure theory and helped extend the concept of fractional dimensions to geometric patterns in nature. Fractal geometry images summarize beauty, yet they can also appear complex and infinitely intricate. In 1892, Mandelbrot defined a fractal as a proper definition of the set of Hausdorff dimension. Later, he expanded on this definition, nothing that fractals are composed of parts that are similar to the whole object.

Fractal geometry and chaos theory offer us a new way to view the world [2]. For many years, basic building blocks have been used to understand the objects around us. Chaos science employs a different geometry known as fractal geometry. This emerging language is used to describe, model, and analyze complex forms found in nature. It has made significant contributions to understanding the growth of cells in various processes, such as cancer cell growth, blood cell formation, chemical carbon bonding, aggregation, cluster formation and dendritic growth. Fractal geometry is the formal study of self-similar structures and is conceptually central to understanding nature's complexity; the fractal dimension provides a measure of structure's complexity. The measured dimension and the box counting dimension methods offer procedures that can be used to measure the fractal dimension of natural shapes such as coastline and mountain ridges, as well as man-made shapes like Frank Lloyd Wright's stained glass windows.

Natural rhythms or harmonic waves are also fractal. Fractal geometry serves as one of the mathematical tools in the science of chaos used to study phenomena that appear chaotic from the perspective of Euclidean geometry and linear mathematics [3]. In biological examples, fractal geometry has been applied to describe, the branching systems in lung airways and the backbone structures of proteins as well as their surface irregularities. Fractals geometry is rough and infinitely complex to visualize. However, it encompasses the process of creating shapes, measuring them and defining them. It characterizes the complexity of image analysis and is particularly useful for qualifying morphologies that are considered random or irregular. Additionally, it provides insights into pattern formation in diffusion and percolation. Typically, fractal dimensions are non-integer dimension and greater than topological dimensions and less than Euclidean dimensions.

DOI: 10.1201/9781003481096-3

3.2 FRACTAL GEOMETRY

A geometric shape with intricate structure at arbitrarily small sizes is known as a fractal in mathematics. Typically, its fractal dimension strictly exceeds its topological dimension. Numerous fractals exhibit similarities at different scales, as demonstrated by the progressively magnified Mandelbrot set. Self-similarity, often referred to as expanding symmetry or unfolding symmetry, is the display of similar patterns at progressively smaller scales. If this reproduction is accurate at all scales, as in the Menger sponge, the shape is referred to as affine self-comparable. Measure theory is an area of mathematics that includes fractal geometry.

Chaos theory and fractal geometry provide a fresh way of looking at the world. Fractal geometry is a distinct kind of geometry used in chaos science. A new language called fractal geometry is being utilized to represent, explain, and examine intricate shapes that observed in nature. Understanding the proliferation of cells in various processes, including aggregation, cluster formation, and dendritic growth, has been greatly aided by fractal geometry. The formal study of self-similar structures is known as fractal geometry, which forms the conceptual basis for comprehending the complexity of nature. An indicator of a structure's complexity is the fractal dimension.

3.3 FRACTALS

The scaling behaviour of fractals sets them apart from finite geometric shapes. For instance, a filled polygon's area is multiplied by four when its edge lengths are doubled. This multiplication is equal to two times the new side length raised to the power of two, which corresponds to the filled polygon's conventional dimension. Similarly, a filled sphere's volume scales by eight if its radius doubles which is equal to two (the ratio of the new to the old radius) raised to the power of three (the filled sphere's conventional dimension). In contrast, the spatial content of a fractal scales by a power that is generally greater than its conventional dimension if all of its one-dimensional lengths are doubled. To distinguish it from the conventional dimension (officially known as the topological dimension), this power is termed the fractal dimension of the geometric object.

Many fractals are nowhere differentiable analytically. Although an infinite fractal curve is still topologically one-dimensional, it can be thought of as winding through space differently than an ordinary line because, it occupies space locally more effectively than an ordinary line, as suggested by its fractal dimension. A line segment is not very different from a proper section of itself. Beginning with ideas of recursion in the 17th century, fractals have progressed through increasingly rigorous mathematical treatment to the study of continuous but non-differentiable functions in the 19th century thanks to the ground-breaking work of Karl Weierstrass's, Bernhard Riemann, and Bernard Bolzano. In the 20th century, the term "fractal" was coined, and interest in fractals and computer-based modeling began to grow.

Mathematicians debate the formal definition of what constitutes a fractal. According to Mandelbrot, it is "beautiful, damn hard, and increasingly useful." Formally speaking, a fractal is a set for which the Hausdorff–Besicovitch dimension strictly surpasses the topological dimension, Later, he expanded and simplified the description to read as follows: "A fractal is a rough or fragmented geometric shape that can be split into parts, each of which is (at least approximately) a reduced-size copy of the whole." He felt that this was too restrictive. Later still, Mandelbrot suggested "using fractal dimension as a general term that applies to all the variants, and using fractal without a pedantic definition.

3.4 FRACTAL PATTERN

Mathematicians agree that theoretical fractals are intricate, indefinitely self-similar mathematical constructions, several instances of which have been developed and examined. Fractals are not only seen in geometric patterns; they can also represent temporal processes. Fractal patterns can be discovered in nature, technology art, and architecture, and have been depicted or researched in visual, tangible, and auditory media to varying degrees of self-similarity. The geometric representations of most chaotic processes include fractals, which are particularly relevant to chaos theory (usually appearing either as attractors or as boundaries between basins of attraction).

The Latin word Fractus, which means "broken" or "fractured," is where mathematician Beniot Mandelbrot first introduced the term "fractal" in 1975. He used it to expand the idea of theoretical fractional dimensions to geometric patterns found in nature. For the general population, who are more likely to be familiar with fractal art than the mathematical concept, the word "fractal" frequently carries different meanings than it does for mathematicians. Even mathematicians find it difficult to define mathematical concepts formally; however anyone with a basic understanding of mathematics can grasp the main ideas.

A simple way to understand the trait of "self-similarity" is to compare it to the process of zooming in with a lens or other equipment on digital photos to reveal finer, previously unseen, structures. However, when this is applied to fractals, nothing changes and the same pattern appears—or, in the case of some fractals, almost identically repeats itself. Self-similarity in and of itself need not be counterintuitive (for example, self-similarity has been considered informally in concepts such as the homunculus, the tiny man inside the mind of the little man inside the head, and the endless regress in parallel mirrors). Fractals differ in that the pattern being replicated must be precise.

3.5 TOPOLOGICAL DIMENSION

The concept of detail is related to another aspect that is easily understood by those with little mathematical background: For example, a fractal with a fractal dimension larger than its topological dimension scales differently than how geometric shapes are typically viewed. A straight line is usually considered one-dimensional; if this line is divided into three equal pieces, each one-third the length of the original, the result will always be three equal pieces. It is commonly accepted that a solid square has two dimensions. If this square is divided into parts, each scaled down by a factor of 1/3 in both dimensions, the result is $3^2 = 9$ pieces.

It is observed that ordinary self-similar objects that are n-dimensional have a total of rn pieces when they are divided into parts scaled down by a factor of 1/r. Let's now examine the Koch curve. It can be divided it into four smaller copies, each with a scale factor of 1/3. Thus, by strict analogy, the only real integer D that satisfies $3^D = 4$ can be considered the "dimension" of the Koch curve. This number is not the dimension of a curve in the conventional sense; rather, it is known as the fractal dimension of the Koch curve. One of the main characteristics of fractals in general is that their dimension differs from the dimension that is commonly recognized (officially called the topological dimension).

This also leads to an understanding of a third feature: fractals as mathematical entities are "nowhere differentiable". This essentially means that standard techniques cannot be used to quantify fractals. To illustrate, imagine trying to find the length of a wavy non-fractal curve by using straight segments of a measuring tool short enough to lie end to end over the waves. By using a tape measure, one might claim that the segments now conform to the curve in the traditional sense. However, the jagged pattern would continue to recur at arbitrarily tiny sizes, making it impossible to find a small enough straight piece to suit an indefinitely "wiggly" fractal curve, such as the Koch snowflake. This implies that an increasing length of the curve would be measured each time the tape measure was used to attempt to fit it more tightly. Because an infinite length of tape is needed to completely trace the curve, the snowflake has an infinite perimeter.

A Koch snowflake is a fractal that begins with an equilateral triangle and replaces the middle third of every line segment with a pair of line segments that form an equilateral bump resembling a Cantor (ternary) set. A common theme in traditional African architecture is the use of fractal scaling, where smaller parts of the structure tend to resemble larger ones, such as a circular village made up of circular houses. The history of fractals traces a path from primarily theoretical studies to modern applications in computer graphics, with several notable figures contributing canonical fractal forms along the way. According to Pick over, the mathematics behind fractals began to take shape in the 17th century when the mathematician and philosopher Gottfried Leibniz contemplated recursive self-similarity (although he mistakenly believed that only the straight line exhibited self-similarity in this sense).

Leibniz used the term "fractional exponents" in his publications, lamenting that "Geometry" was unaware of them at the time. Several historical accounts state that following this period, few mathematicians attempted address these problems, and the work of those who did remained largely hidden due to opposition to such strange new ideas, which were often referred to as mathematical "monsters." It was not until two centuries later, on July 18, 1872, at the Royal Prussian Academy of Sciences, that Karl Weierstrass's provided the first definition of a function with a graph that is still regarded as a fractal today, characterized by the counterintuitive property of being everywhere continuous but nowhere differentiable. Furthermore, as the summation index rises, the quotient difference becomes arbitrarily large. Shortly thereafter, in 1883, Georg Cantor, who had attended Weierstrass's lectures, published instances of what are now known as fractals—subsets of the real line with peculiar features. In the latter half of that country, Henri Poincaré and Felix Klein also devised a class of fractals known as "self-inverse.

3.6 BOX COUNTING DIMENSION

The minimum number of sets of diameters δ that can cover the set F is denoted by a numeral $N_\delta F$, showing the quantity of clusters of diameter related to δ into which it can be partitioned. The dimension of F expresses the process in which $N_\delta F$ develops as $\delta \to 0$. If $N_\delta F$ yields at least approximately a power law $N_\delta F \simeq C\delta^{-S}$, we can define a positive constant C and S, stating that F has a box dimensions. Box counting is a technique for collecting data for the analysis of complex pattern by dividing a dataset, object, image, etc. into progressively smaller parts—typically in the form of a "box"—and analyzing the pieces at each smaller size. The methodology can be thought of as essentially zooming in or out using optical or computer-based techniques to study how scale affects detail observations. However, with box counting, the investigator modifies the size of the element used to examine the object or pattern rather than altering the magnification or resolution of a lens.

Patterns in one, two, and three dimensions have been subjected to computer-based box counting techniques. While the basic method can be used to physically study certain patterns, it is typically adapted in software for use on patterns derived from digital sources. The method originated with fractal analysis and is still applied today proving useful in related topics such as multifractal analysis and lacunarity.

Depending on the topic and type of research being conducted, different features are obtained during box counting. For example, binary data—that is, data with only two colors, typically black and white—and grayscale digital images such as JPEGs, TIFFs and the like—are two well-researched topics in box counting. When box counting is applied to patterns derived from these still images, the raw data captured is usually based on pixel attributes, such as a predefined color value or a range of colors or intensities. The information regarding whether the box contained any pixels of the predefined color or range is often recorded as either yes or no when box counting is used to calculate a fractal dimension known as the box counting dimension.

3.7 FRACTAL DIMENSION

Fractal dimensions can be beneficial in machine learning (ML) for handling data and dimensionality reduction. They serve as useful tools for various types of technical evaluations, as they represent ratios of a figure's complexity at different scales. For example, fractal dimension is frequently applied to dimensionality reduction, a ML challenge that involves simplifying data set analysis so that the system can generate a different model with fewer parameters. Two methods for applying dimensionality reduction feature extraction and feature selection modify the model based on user requirements. Fractal dimensions are generally useful in illustrating how scaling affects a model or modeled object. Consider graphing a highly complicated shape at a certain scale, and then reducing the scale; the data points become smaller and more convergent. This is the type of task that fractal dimensions can measure and evaluate.

3.7.1 Points of Increase

A real-valued function f has a global point of increase within the interval (a, b) if there exists a view point $t_0 \in f(a,b)$ such that

$$f(t) \le f(t_0) \text{ for all } t \in (a,t_0) \text{ and } f(t_0) \le f(t) \text{ for all } t \in (t_0,b). \tag{3.1}$$

We say t_0 is a neighborhood element of increase if it is a global point of increase in several intervals. Here we show that the no expansion phenomenon holds for arbitrary symmetric random walks and may thus be considered as a combinatorial outcome of changes in arbitrary sums.

3.7.2 The Hausdorff Measure

The Hausdorff dimension of a self-affine set should be either 0 or ∞. Suppose the integer set D has non-uniform parallel fibres and Let $\gamma = \dim$ (K (D)), Then $H^{\gamma}(K(D)) = \infty$

In addition, K(D) is not always s-bounded for H^{γ}. Since the intersections of K (D) with the D square shapes of the original are interpreted in relation to one another and t. The Hausdorff measure H^{γ} is interpretation invariant; its limitation to K (D) must allocate each of these square shapes a similar measure.

3.7.3 Topology of Uniform Convergence

Let X be set and (Y, V) a uniform space. For each $V \in v$ let

$$B(V) = \{(f,g) : f(x), g(x)) \in V \; \forall x \in X\} \tag{3.2}$$

Then $\{B(V) : V \in v\}$ is a base for uniformity $\mathcal{M}$ on Y^x. this uniformity $\mathcal{M}$ is called the uniformity of uniform the uniforms convergence or uniform convergence uniformity. The topology $\mathfrak{I}_p$ of $\mathcal{M}$ is represented by the topology of uniform convergence or uniform convergence topology.

3.8 SIMULATED FRACTALS

Since physical time and space have practical bounds, fractal patterns have been modeled extensively, albeit within a range of scales rather than infinitely. Fractal properties in actual occurrences or theoretical fractals can be simulated via models. The results of this modeling process can serve as standards for fractal analysis, highly creative depictions, or outputs for further research. A list of specific technological uses for fractals can be found elsewhere. The term "Fractals" is often used to describe images and other modeling outputs, even when they lack true fractal features—for instance, when zooming in on a portion of a fractal image that exhibits no fractal attributes- for instance, when zooming in on a portion of fractal image that exhibits attributes. These outputs may also contain display or computation artifacts that are not characteristic of real fractals.

Modeled fractals can manifest digital graphics, sounds, electrochemical patterns, rhythms of the circadian cycle, and more. In a process commonly referred to as "in silico" modeling, fractal patterns have been electronically and physically recreated in three dimensions. Fractal models are typically generated using software that employs methods akin to those mentioned above. For example, recursive algorithms and L-systems techniques can simulate branching patterns found in nature, such as those observed in trees, ferns, nervous system cells, and blood and lung vasculature. A branch from a tree or a frond from a fern serves as a small reproduction of the whole; it is not identical, but is similar in nature. Recursive patterns can be seen in these instances. Similarly, numerous irregular real-world features, such as mountains and coastlines, have been described or created using random fractals. One drawback of fractal modeling is that a model's similarity to a natural phenomenon does not imply that the phenomenon was produced through processes similar to those employed in the modeling methods.

3.9 FRACTAL GEOMETRY WITH ARCHITECTURE

Fractal geometry has been utilized in architecture in various ways due to its complex patterns and self-similarity across different scales.

3.9.1 Form Generation

By applying fractal principles, architects can create intricate and captivating forms. Fractal patterns can influence designs for exteriors, interiors, and entire buildings. Because fractals are recursive, architects can generate intricate yet cohesive designs with repeating patterns at varying scales. Structural Efficiency: High structural often exhibit high structural efficiencies. By integrating fractal geometry into their structural systems, architects can develop buildings that are both visually appealing and structurally sound. Fractal-based constructions can distribute loads more evenly, potentially utilizing fewer resources while maintaining strength and stability.

3.9.2 Space Optimization

Buildings can make the best use of their available space by utilizing fractal geometry. By arranging rooms, hallways, and other architectural components in fractal patterns, architects can create spatial layouts that enhance both efficiency and functionality.

3.9.3 Biophilic Design

Fractals are evident in many natural phenomena, including clouds, coastlines, and tree branching patterns. When incorporated into architectural design, fractal geometry can evoke feelings of connectedness to nature, aligning with the principles of biophilic design. Fractal designs not only add visual interest but also foster a sense of harmony with nature in buildings, potentially enhancing tenant productivity and well-being.

3.9.4 Building Facade Design

Using fractal geometry, architectural facades can be crafted with visually striking patterns that serve functional purposes such as ventilation and shade. By manipulating fractal patterns, architects can regulate the amount of heat and light entering structure thereby improving occupant comfort and energy efficiency.

3.9.5 Urban Planning

Applying fractal principles can lead to the design of cities and neighbourhoods more aesthetically pleasing, environmentally friendly, and efficient. The integration of fractal patterns into the design of parks, streets, and public spaces enables urban planners to create landscapes that are visually appealing and coherent, while still accommodating a variety of functions. Overall fractal geometry, offers provides architects and urban planners a versatile toolkit for creating unique, beautiful, and functional designs that reflect the complexity and interconnectivity of the natural world.

3.10 FRACTAL GEOMETRY WITH IMAGE PROCESSING

Image processing greatly benefits from the use of fractal geometry, which provides several methods for producing, evaluating, and compressing images. Here is how it is applied:

3.10.1 Fractal Image Compression

Fractal image compression is a method that utilizes the self-similarity characteristics of fractal for picture compression. Instead of saving individual pixels, fractal compression techniques store mathematical descriptions of self-similar patterns within the image [4]. By encoding a fractal transformation that converts one area of the image to another, it allows for the representation of large images with comparatively little data. Fractal compression particularly useful in situations where storage or bandwidth is constrained, as it can achieve high compression ratios without sacrificing image quality.

3.10.2 Image Synthesis

Fractal geometry can be used to artificially create images with intricate and visually appealing. Fractal algorithms, such as the collage theorem or the iterated function system (IFS), can generate complex patterns that mimic textures, clouds, and other natural phenomena. These generated images can be employed in computer graphics, digital art, and special effects to create realistic or creative visualizations.

3.10.3 Image Analysis

Fractal analysis tools can measure the self-similarity and complexity of images. By quantifying an image's space-filling characteristics, fractal dimension estimates provide insights into its structural complexity. Fractal-based texture analysis enables applications in object detection, image segmentation, and medical imaging by categorizing and describing textures based on their fractal characteristics [5].

3.10.4 Image Enhancement

Fractal techniques can improve the quality of digital photographs. Fractal interpolation algorithms can repair or replace damaged or missing areas of an image by extrapolating from nearby pixels, resulting in visually appealing outcomes. Additionally, fractal-based denoising methods utilize self-similarity to eliminate noise while preserving important aspects of the image, thereby enhancing quality for application in satellite or medical imaging.

3.10.5 Image Registration

Image registration involves aligning multiple photographs of the same scene taken at different times or from various views. Fractal features can facilitate this process. Fractal-based registration techniques simplify the alignment of images by identifying comparable structures or patterns, making them useful for applications such as panorama stitching, remote sensing, or medical image registration [6].

Overall, fractal geometry offers a rich framework for digital image analysis, processing, and synthesis. It provides solutions for image synthesis, compression, analysis, enhancement, and registration across a variety of fields, from computer graphics to medical imaging.

3.11 FRACTAL GEOMETRY IN ARCHITECTURE

Fractal geometry has fascinating applications in the film industry, influencing both the technical aspects of filming and the visual aesthetics of films:

3.11.1 Visual Effects

Filmmakers frequently utilize fractal geometry to create striking visual effects. Fractal patterns can simulate textures, clouds, and other natural phenomena, giving (computer-generated imagery) images a realistic and captivating appearance. Fractal algorithms are essential to modern visual effects, whether used to craft complex landscapes for fantasy worlds or to mimic natural phenomena like fire or water.

3.11.2 Digital Matte Painting

This approach allows filmmakers to create backgrounds and environments that would be too costly, challenging or impossible to film in real life. When producing realistic and detailed matte paintings of large-scale landscapes, intricate cityscapes, or other complex locations, fractal geometry serves as a valuable tool. Fractal-based algorithms can generate natural-looking landscapes, forests, mountains, and architectural structures, enriching the background of film sequences.

3.11.3 Procedural Generation

Procedural generation techniques employ fractal algorithms to build expansive and varied virtual environments for movies. By specifying rules and constraints, filmmakers can create entire landscapes, towns, or civilizations procedurally. This method enables the generation of vast, immersive environments with minimal manual intervention. Thanks to fractal-based procedural generation, filmmakers can explore extensive fictional worlds and design unique settings tailored to their storytelling needs.

3.11.4 Title Sequences and Graphics

Fractal geometry also influences the creation of visual themes, title sequences, and graphics films. Fractal forms and patterns can enhance on-screen graphics with greater visual appeal, depth, and complexity, improving a movie's overall aesthetic. Whether constructing, dynamic title sequences or incorporating fractal elements into character designs or motion graphics, filmmakers can leverage fractal geometry as a versatile tool for producing visually captivating visuals.

3.11.5 Experimental Film Making

The abstract and visually striking qualities of fractal geometry have also inspired experimental filmmakers. Some directors use fractal patterns and structures as the foundation for their works, employing them to convey a sense of infinity, intricacy, or spirituality or to produce captivating abstract imagery. Experimental films featuring fractal themes can subvert traditionally narratives and engage audiences through their visual and emotional resonance. In summary, fractal geometry enhances the film industry by providing filmmakers with the means to create visually spectacular effects, construct vast virtual worlds, develop aesthetically appealing graphics, and explore abstract visual themes. It finds applications across a wide range of filmmaking contexts, from production design and visual effects to experimental and visual narratives.

3.12 FRACTAL GEOMETRY IN THE BRAIN

When used to investigate the brain, fractal geometry has intriguing ramifications that shed light on its dynamics, structure, and function:

3.12.1 Brain Structure

The intricate and self-similar structure of the brain can be described at various scales using fractal geometry. Neuronal arrangements, blood vessel patterns, and cortical folding patterns can all be observed as fractal patterns in brain pictures obtained from magnetic resonance imaging (MRI) scans. These fractal features reveal important details about the structure and connectivity of the brain, illuminating its complexity and efficiency.

3.12.2 Brain Dynamics

The activity of the brain exhibits fractal characteristics at different temporal scales. Fractal patterns in the fluctuations of cerebral activity, as found electroencephalography or data suggest the presence of self-similarity and long-range correlations. These fractal dynamics are believed to play a role in brain synchronization, information processing, and cognitive function.

3.12.3 Brain Function

The functional organization of the brain and its cognitive processes can also be understood through the lens of fractal geometry. Fractal patterns in the synchronization of neural activity across various brain regions can be identified through fractal analysis of functional brain networks, derived from methods such as FMRI or EEG connectivity analysis. These fractal features may reflect the brain's ability to flexibly adjust its functional connections, supporting complex cognitive processes like memory, learning, and decision-making.

3.12.4 Brain Pathology

Fractal analysis may benefit the study of brain pathologies and disorders. Conditions including Alzheimer's disease, epilepsy, schizophrenia, and traumatic brain injury have all been linked to changes in the fractal patterns of brain structure or dynamics. Characterizing these fractal aberrations could aid in understanding causes of neurological and mental illnesses as well as formulating diagnostic and treatment plans.

3.12.5 Brain-Inspired Computing

Fractal principles are also influencing new ideas in artificial intelligence and brain-inspired computing. Neural networks, learning algorithms, and computational architectures can be designed using fractal algorithms and models to mimic the self-organizing, adaptive, and hierarchical characteristics of the brain. The goal of using fractal geometry in computer models of brain activity is to create artificial intelligence systems that are more reliable and efficient.

In conclusion, fractal geometry provides an effective framework for exploring the dynamics, structure, and function of the brain, offering valuable insights into its complexity and organization at various scales. Researchers can deepen their understanding of brain function, pathology, and the fundamentals of cognition and consciousness by applying fractal analysis to brain imaging data and computational models.

3.13 FRACTAL GEOMETRY IN DESIGNING CLOTHES

Fractal geometry can inspire and impact clothing design, in several ways, offering designer opportunities to produce distinctive and eye-catching apparel:

Textile Designs: Thanks to fractal geometry, designers have countless opportunities to create complex and visually appealing textile designs. Fractal-based motifs—such as self-similar geometric patterns or organic fractal structures—can be incorporated into fabric designs through printing, weaving, or embroidery process. These fractal patterns enhance the overall aesthetic appeal of clothing by adding depth, intricacy, and visual intrigue to textiles.

Silhouette and Structure: Fractal geometry can influence the silhouette and structure of clothing designs. By using fractal-inspired shapes—such as spirals, branching patterns, fractal curves designers can create innovative clothing style that play with repetition, symmetry, and asymmetry.

Introducing fractal components into the structure of garments allows designers to give their creations dynamic and sculptural features, resulting in visually remarkable and unique pieces, Texture and Surface Design: Designers can gain considerable inspiration from fractal geometry when producing tactile and textured clothing surfaces. Techniques such as laser cutting, 3D printing, or embossing can be utilized to create fractal-based textures including complex fractal reliefs and rough surfaces, These textured elements enhance the wearer's experience by adding depth, dimension, and sensory intrigue to clothing. Fractal geometry can influence pattern-cutting and draping methods in clothing creation. Concepts such as scaling and self-similarity can inspire innovative approaches to altering patterns and constructing garments. By integrating fractal principles to their design processes, designers can produce clothing with distinctive shapes, folds, and seam lines that embody the fundamental ideas of fractal geometry. Fractal geometry has the potential to inspire designers to explore novel color and gradient effects in their apparel designs, Fractal based color schemes can be employed to create visually striking and harmonious compositions. Examples include fractal gradients and color palettes derived from natural fractal patterns. Through the application of digital printing technologies and color blending techniques, designers can achieve captivating color effects that reflect the complexity and elegance of fractal geometry.

All things considered, fractal geometry provides fashion designers with a plethora of imaginative options for creating outfits that are striking appearance, intellectually complex, and technologically avant-garde. By drawing inspiration from fractal patterns, shapes, and principles, fashion designers can create clothing that defies convention and captures the attention of both the wearer and onlookers.

3.14 CONCLUSION

In summary, the study of fractals and fractal geometry has fundamentally changed our perception of mathematical and natural structures by illuminating the intricacy and self-similarity of the world around us. Thanks to fractal, we now have a robust framework for describing asymmetrical and fractured shapes—forms that defy conventional Euclidean geometry. Applications of fractals can be found in many fields, including biology, physics, computer science, and the arts. They provide a means to accurately describe intricate phenomena like vascular networks, clouds, and coastlines while exposing underlying patterns and structures that were previously obscured. Additionally, fractal geometry has opened up new avenues for research and technological advancement giving rise to fractal-based algorithms for data analysis, signal processing, and image compression, resulting in more reliable and effective solutions across various fields.

Despite these developments, the study of fractals still faces challenges and unsolved issues. The dynamics of fractal systems, their behaviour at various scales and conditions, and their implications for real-world phenomena further exploration. To realize the fractal geometry's full potential of fractal geometry and expand its applications, collaboration among mathematicians, physicists, engineers, and artists is encouraged due to its interdisciplinary nature. In short, the study of fractals and fractal geometry remain a topic great interest to scholars and enthusiasts alike providing countless opportunities for exploration, creativity, and innovation in the quest to comprehend the intricate details of the natural world and the mathematical principles that underpin it.

REFERENCES

1. Garg, A., A review on natural phenomenon of fractal geometry, *International Journal of Computer Applications*, Vol. 975, p. 8887 (2014).
2. Barnsely, M. F., *Fractals Everywhere*, 2nd ed., Academic Press, Boston, MA (1993).
3. Mandelbrot, B., *The Fractal Geometry of Nature*, Vol. 1, WH Freeman, New York (1982).
4. Falconer, K. J., *The Geometry of Fractal Sets*, Cambridge University Press, Cambridge (1985).
5. Kigmi, J., *Analysis on Fractals*, Cambridge University Press, Cambridge (2011).
6. Falconer, K. J., *Techniques in the Fractal Geometry*, John Wiley & Sons, Ltd, Chichester (1997).

4 Classical Iterated Function System

4.1 INTRODUCTION

Many peoples are fascinated by the appearance and patterns of fractals. The term "Fractals" was coined by Beniot Mandelbrot in 1975 who dedicated many years to the development of this concept. Today, the study of fractals plays vital role of solving real world problems. Fractals studies based on their inherent properties, find applications in general mathematics, computer simulations, imaging and graphical processing. This chapter analyzes different types of fractals available. Iterated function systems (IFSs) are a mathematical construction technique that produces fractals, many of which are self-similar [1]. Set theory is more closely related to IFS fractals than it is to fractal geometry. In 1981, they were first presented. In mathematics, an iterated function is one that is applied to it two or more times. Iteration is the process of applying a function repeatedly, where the result of one step is input for the next. For instance, the following steps could make up a basic algorithm for eating cereal in the morning: Pour cereal into a bowl, pour some milk over your cereal, and repeat as necessary. Iteration can either be definite iteration, in which the total number of repetitions is predetermined, or indefinite, where code blocks are executed indefinitely until a condition is satisfied. This method is sometimes is referred to as evolutionary or circular development. A single iteration of the sequence completes all steps in the prescribed order, creating a loop when a set of instructions is carried out repeatedly. The difference between iteration and recursion is important. Recursion is a technique where the function breaks a problem into smaller, easier-to-manage sub-problems by calling itself within its body. While, iteration is a method that runs a code block repeatedly until a specific condition is satisfied.

Similar to agile processed the iterative approach" is common in the engineering field, but it is also used across various fields. The iterative approach lowers risk, improves efficiency and provides a more adaptable and dynamic approach to problem-solving. Iteration is achieved through "loops "which can be either count-controlled or condition controlled.

Furthermore, IFS-based methods make it easier to compress digital images by using the self-similarity inherent in fractal patterns, which helps to accomplish effective data representation and transmission. Classical IFS provide a forum for artistic expression and research beyond their technical uses. Using the generative potential of IFS algorithms, artists explore the balance between symmetry and complexity, and order and chaos, and creating visually stunning creations [2]. IFS-generated graphics have a fractal beauty that makes one think about the intricate details of mathematical abstraction and the underlying architecture of the natural world. To sum up, traditional IFS provide a blend of computing effectiveness, mathematical elegance,

DOI: 10.1201/9781003481096-4

and creative flair [3]. Their influence extends beyond academic fields, providing insights into of the dynamics of iterative processes, the geometry of fractal formations, and the aesthetics of visual art.

4.2 BENEFITS OF RECURSION

Compared to iterative code, recursive code can be easier to read and comprehend especially for certain algorithms and data structures, where recursion is necessary. Recursion allows shortening code, making it easier for programmers and users to read and comprehend. Iterative processes enable rapid refinement and revision of products. This approach is effective because one can develop a product incrementally rather than starting over from scratch whenever a change is needed. Iterative design allows for rapid concept generation and testing. Ideas that show potential can be immediately iterated until they become fully developed, while those that don't can be promptly dropped.

The process of creating, prototyping, and testing various iterations of a product in successive cycles is known as iterative development. This differs from the more linear approach typically used in product development. In programming, iteration is specified by using for, while, and repeat statements. Before each repetition, the final value is compared to the initial value of the index variable in the statement. Iterative structures, also known as repetitive control structures, are groups of code designed to repeat a series of linked statements. Repetition, also known as iteration, can occur zero or more times until stopping until a control value or condition is satisfied [4].

4.3 DIMENSION OF A FRACTAL GRAPH

A fractal dimension is used to determine the complexity of a system based on its scale. It provides a ration that compares a statistical indexes of complexity with the detail in its pattern. While fractal geometry has a long history in mathematics, the term fractured dimension was introduced by Beniot Mandelbrot in 1967, based on the concept of self-similarity. He was inspired by Lewis Richardson's work of measuring Britain coastline, which illustrated the counterintuitive nature of length when measured with different scales. Many mathematical definition of fractal dimension have been developed, all focusing on how details change as he scale changes. An object that looks similar at different scales is described as self-similar. Fractals are collections of such self-similar objects. The complexity of these objects is measured using power law [5].

Fractals are collections of such self-similar object. The complexity of these objects is measured by using power law: $N = s^d$, where $d = \frac{\log N}{\log s}$ is the dimension of scale law with parameter N and s. This measurement is called as Hausdorff dimension.

4.4 FEATURES

The word fractal comes from the Latin word "Fractus" meaning broken or subdivided or fragmented. In 1918, Polish-born mathematician Felix Hausdorff introduced the concept of complex geometric shapes that have fractional dimensions. A fractal exhibits the following characteristics:

- It forms well-structured framework at proportions.
- It cannot be adequately described using traditional Euclidean geometric language.
- It has self-similar property which can be exact, approximate, or stochastic.
- The Hausdorff dimension measures the roughness or complexity of shapes, often exceeding one dimension.
- It has clear and recurrent definition.

Not all self-similarity objects are considered fractals. For example, a straight Euclidean line has accurate self-similarity but fails to exhibit the characteristics of fractals, and can be fully described in the Euclidean terms.

4.5 CLASSIFICATION OF SELF–SIMILARITY

Fractals can be classified into three categories according to their characteristics and degree of similarity. Self-similar objects are either accurately or partially similar themselves meaning the whole object has the same shape or some parts are similar to the whole.

4.5.1 Accurate Self-Similarity in Fractal Graphs

This is the most prominent type of self-similarity. In these fractals, objects look identical at various scales. Fractals defined by IFS often display self-similarity accurately. Scale invariance is one of the key properties of self-similarity meaning a small part of the object at any magnification is similar to the whole of the body of the object. Many mathematical fractals display accurate self-similarity such as Cantor set, Von Koch Curve, Koch snowflake, and Sierpenski arrow head curve. These fractals exhibit are both exact self-similarity and scale invariance meaning they can be continuously magnified without changing their shape. Many self-similar fractals are found in the world. Some examples are given in the following paragraphs [6].

4.5.1.1 Cantor Set

Cantor set is a basic example of a fractal graph, starting from a topological dimension meaning it begins with a unit interval [0, 1]. Henry John Smith first introduces the concept; the set became famous through the work of George cantor, after whom it is named. Initially, the Cantor set has only two vertices and one edge. It is evaporated by subsidiary rule. The single edge divided into three equal parts and the middle third eliminated. This results in four non-adjacent edges and eight vertices. This process is called as first iteration. The same process is applied each edge subsequent iteration. It is an example of string fractals and is widely recognized as a compact set. Figure 4.1 shows the Cantor set.

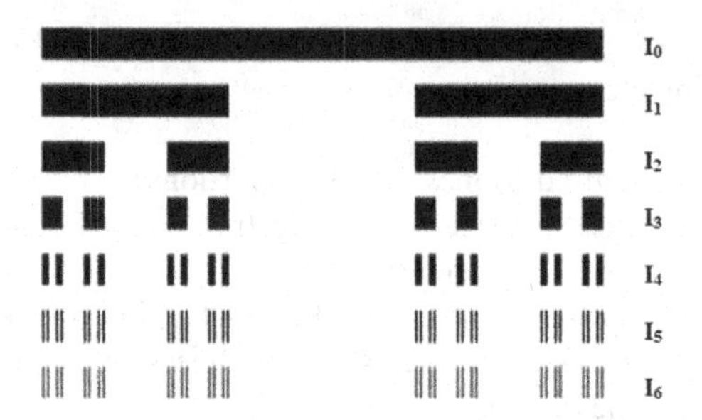

FIGURE 4.1 Cantor set.

4.5.1.2 Fractal Dimension of the Cantor Set

Cantor Set has examined carefully in which has the portion of the object corresponding to one-third of the original segment. It resembles that it is an exact repetition of complete object with 1/3rd of its scale. Hence, if Hausdorff dimension is d then $3^d = 2$, and $d = \frac{\log 2}{\log 3} = 0.631$, a non-intrinsic quantity. The topological dimension of the Cantor set is zero.

4.5.2 von Koch Curve

In 1904, mathematician Hedge von koch, introduced the Von Koch curve which is an example of exact self-similarity. He studied under Gosta and became a professor at stockholn University. It was used in Mandelbrot's work on fractal graphs. The initial iteration of the curve begins with a unit interval straight line. In the first iteration, the segment is partitioned into three equal parts. The middle part is replaced by two parts, forming an equilateral triangle without a base. After the first iteration, the curve consists of four segments with its total length increasing by 1.33. This process is repeated and the resulting four segments are increased in the following iterations.

4.5.3 Fractal Dimension of the Koch Curve

A fractal is an object or function that displays self- similarity at various scales in a technical sense. For an object with dimension DM and it can be subdivided into N equal part. Each of which is smaller in size. Let r represent the length of the curve. Then

$$r = \frac{1}{N\left(\frac{1}{DM}\right)} \tag{4.1}$$

$$DM = \frac{\log\,(N)}{\log(1/r)} \tag{4.2}$$

Let I_0 be an unit length. I_1 has the four segments obtained by removing the middle third of I_0. Let $N = 4$. Each line segment derived from its parent segment and is one-third of its whole. The Von Koch Curve has equal edges in all iterations. The main measurement tool in fractal geometry is dimension which comes in

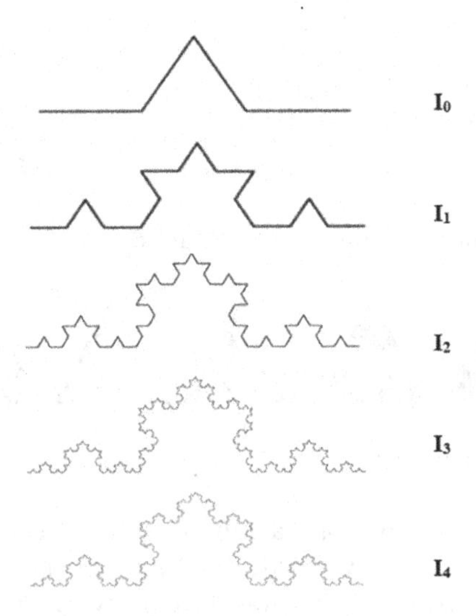

FIGURE 4.2 von Koch curve.

many forms. For example, a smooth curve has one-dimension while a surface has two-dimension.

The Cantor set has a dimension of $DM = \frac{\log 2}{\log 3} = 0.631$.

The von Koch curve has a dimension of $DM = \frac{\log 4}{\log 3} = 1.262$.

Figure 4.2 is an image of the von Koch curve.

4.5.4 Koch Snowflake

The Koch Snowflake is another example of a self-similar fractal curve and is one of the oldest fractals that have been described. In 1904, Hedge von koch presented a paper titled "On a continuous curve without Tangents, Constructible from Elementary Geometry". In this work, he proved against Weierstrass's function with a self-similarity object. The Koch snowflake is built iteratively through multiple stages. The first stage consists of an equilateral triangle. In the subsequent stages, an outward bending line is added to each side of the triangle from the previous stage. It makes a small size equilateral triangle at the whole shape. The area of the Koch Snowflake increased (8/5) times compared to the area of the initial equilateral triangle. Unlike the area, which remains finite the perimeter becomes infinite. Figure 4.3 depicts a Koch snowflake: Iteration 1,2 & 3.

Figure 4.4 depicts a Koch snowflake: Iteration 4,5 & 6

4.5.5 Sierpinski Triangle

Sierpinski triangle is called the Sierpenski gasket or Sierpenski sieve. It exhibits an exact self-similarity Graph as a graph. Its overall shape is that of an equilateral triangle, which is partitioned into smaller parts. Each part is recursively represented as a smaller equilateral triangle. The gasket is generated by a mathematical pattern

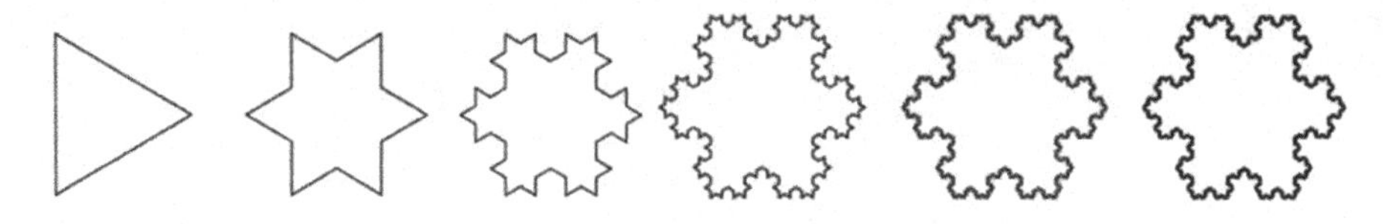

FIGURE 4.3 Koch snowflake: Iteration 1,2,3,4,5 and 6

FIGURE 4.4 Sierpinski triangle.

and is reproducible at any level of magnification or reduction. It was invented by the Polish mathematician Waclaw Sierpenski. The Sierpenski gasket serves various decorative purposes. Sierpinski work was done many centuries ago, long before it became widely recognized. The gasket is formed by repeatedly removing triangular subsets from an equilateral triangle.

4.5.5.1 Fractal Dimension of Sierpinski Triangle

The Sierpinski triangle begins with an equilateral triangle of equal length. By using the midpoint of each, it is divided into four congruent equilateral triangle. Each part will be half the scale of the original. The innermost triangle is then removed, leaving being three triangles. This process is repeated for each triangle, continuing indefinitely. The Hausdorff dimension DM of this object is computed. It is observed that, if a copy of the object is scaled by a factor of 2. Only three copies are needed to cover an original triangle.

Hence, $2^{DM} = 3$ and $DM = \frac{\log 3}{\log 2} = 1.585$.

Figure 4.4 depicts Sierpinski triangle.

4.5.6 Sierpinski Carpet

The Sierpinski carpet is a regular uniform fractal graph is invented by Waclaw Sierpinski at 1916. It follows the technique of subdividing a shape into smaller copies that resemble the original. The process involves removing one or more copies and continuing recursively. This concept can also be extended to modified shapes. Figure 4.5 shows an image of a few iterations of the Sierpinski carpet.

An equilateral triangle is subdivided into four smaller equilateral triangles, with the middle triangle removed. The remaining triangles lead to a recursive pattern those results in the Sierpinski triangle. A similar construction in three dimensions, based on cubes, is known as the Menger sponge. Sierpinski asserted that his carpet is a comprehensive plane curve, which is a solid subset of the plane with a Lebesgue covering dimension. Every subset of the plane with these properties is homeomorphic to a subset of the Sierpinski carpet. Several iterations of the Sierpinski carpet are used in mobile phones and Wi-Fi fractal antenna. This structure is easily accommodates multiple frequencies easily due to self-similarity and scale invariance. The Hausdorff dimension of the sierpinski carpet is $d = \frac{\log 8}{\log 3} = 1.8928$.

FIGURE 4.5 Few iterations of the Sierpinski carpet.

4.5.7 Fractal Antenna

The Fractal Antenna is a self-similar design. It was discovered by Dr.Nathan Cohen in 1988 and is widely accepted around the world. This design offers greater assimilation, adequate length and diversity compared to other antennas available globally. It is used to transfer electromagnetic emissions over its total surface area and is constructed using metamaterials. An antenna utilizing the Hilbert curve serves as a prime example of a fractal antenna structure, which is employed in various real-life electrical applications.

4.5.7.1 Metamaterial

Fractal Antenna can be manufactured using metamaterials. Unlike natural materials such as biotic or inorganic substances, metamaterials are synthesized from multiple components. They are made from a combination of metals and plastics arranged in iterative patterns at various scales smaller than the wavelength of the electromagnetic signals they interact with metamaterials are not designed based on the properties of conventional materials; instead, they exhibit unique properties that enable the creati9on of advanced structural designs. The characteristics of electromagnetic waves are closely linked to the essential properties of metamaterials, including pattern, geometry, assimilation and arrangement. Metamaterials are used for various applications, including insulation, attenuation, enhancement and wave bending to achieve specific benefits.

4.5.7.2 Advantages

- The new development of Fractal antenna is recognized as one of the most advanced wideband and multiband antennas in the world. It reduces the cost of antennas as well as the materials needed for their construction. Typically, antennas meet various wireless needs and are useful in cellular telephones and microwave communication, etc. [7].
- They are manufactured to have minimal size and weight. This small size leads to less cumbersome installation. Additionally, Fractal antenna offer various design flexibilities, enhanced aesthetics and minimal interaction with other antennas.
- They are easy to install due to their lightweight design, which also reduces the need for mechanical support. They are constructed without the use of capacitors [8].
- Traditionally designed antennas require more components for their creation. In contrast, Fractal antennas achieve the same functionality with fewer components, reducing the need for spare parts and lowering material costs.
- Fractal antenna provides better coverage areas while reducing the number of antennas needed.

FIGURE 4.6 Hilbert curve

4.6 HILBERT CURVE

The Hilbert curve is one of the most remarkable structures of Fractal Antennas. It is a highly effective antenna design compared to other models. This structure is characterized by its self-similarity properties. Generally self-similar fractal antenna designs provide enhanced perception, capable magnitude and reduced fluctuation. It outperforms other antennas available in the world and is used to transmit electromagnetic radiation over a limited area. It is manufactured using metamaterials.

The Hilbert curve antenna structure was discovered by the German mathematician David Hilbert. It is a continuous fractal space- filling curve. A space filling curve is a contour that occupies two dimensional spaces. Geometrically, it represents an open path with two-degree regular graph. H_m denoted the m-dimensional Hilbert space. The length of H_m is $2^m - \frac{1}{2^m}$ determined by the number of iterations of the limiting curve. It is enclosed within a square shape with a finite surface area. Undoubtedly Hilbert curve forms 2-dimensional Hausdorff unit square. Figure 4.6 presents an image of the Hilbert curve.

4.7 SIERPINSKI ARROWHEAD CURVE

The Sierpinski arrowhead curve is an example of an open path fractal graph. It is closely related to the Sierpenski triangle which is one of the earliest described and most famous fractal graph in the world. This fractal exhibits an exact self-similarity. The construction begins with a shape resembling half a hexagon. It can be implemented using two alternating rules. The curve consists of horizontal and vertical lines denoted by H and V with direction also being considered. The lines of the graph can be turned left or right at 60^0 degree angles with left and right movements represented as L and R respectively. Each vertical line follows the pattern H R V R H while the horizontal line moved follows into V L H L V. In each iteration, the number of edges increases by a factor of three, which can be expressed as 3^n where n is the number of iterations. Figure 4.7 provides an example of the Sierpinski arrowhead curve.

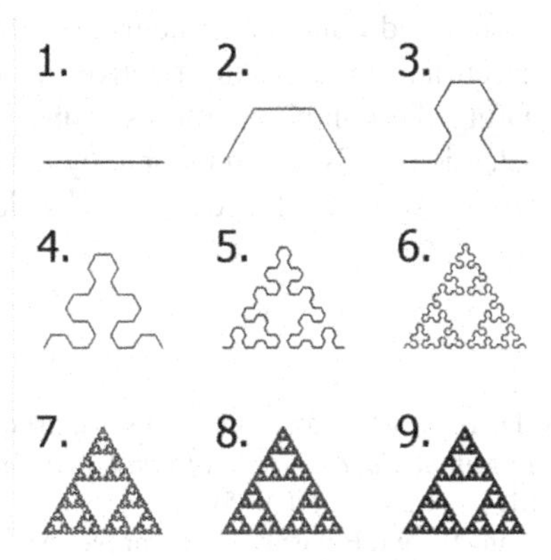

FIGURE 4.7 Example of Sierpinski arrowhead curve.

4.7.1 Vertices and Edges

In Iteration 1, the shape consists of a half hexagon with three edges and four vertices. Each edge is multiplied by three. In the second iteration of Sierpenski arrowhead curve has 9 edges and 10 vertices. In the third iteration, Sierpenski arrowhead curve have 27 edges and 28 vertices. In the fourth iteration, Sierpenski arrowhead curve have 81 edges and 82 vertices. The Sierpenski arrowhead curve follows a common formula for increasing vertices and edges. In general, m-th iteration of the Sierpenski arrowhead curve has $3^m + 1$ vertices and 3^m edges.

4.8 CONCLUSION

To sum up, traditional Iterated Function Systems (IFS) are a fundamental tenet of fractal geometry, offering an adaptable structure for producing and understanding a wide range of fractal patterns and forms. IFSs generate complex structures with fine detail, recursion, and self-similarity at all scales by repeatedly applying basic transformations to an initial set of points or polygons. One of the most striking aspects of classical IFSs their ability to preserve the essence of natural shapes, including ferns, trees, and coastlines. IFSs serve as mathematical languages that encode the geometric qualities of these forms into a set of transformation rules allowing for the exquisite beauty of the natural world to be described and replicated mathematically.

Additionally, traditional IFSs have been essential in bridging the aesthetic and mathematical divides by encouraging designers, artists, and enthusiasts to explore the aesthetic potential of fractal geometry. The creative applications of IFSs, ranging from visually arresting imagery to engaging sculptures are benefited both the artistic community and the mathematical discourse surrounding fractals. Although classical IFSs have been thoroughly examined and utilized, there is remains a need for further investigation and study. Advances in computational methods, algorithmic techniques, and theoretical analysis, are pushing the limits of what is achievable with IFSs, creating new opportunities for research and innovation.

In conclusion, classical Iterated Function Systems are a fundamental component of fractal geometry research and provide an effective means of simulating, comprehending, and appreciating the complex patterns found in both the natural and mathematical domains. Classical IFSs will undoubtedly continue to be an essential tool in our pursuit of creativity and knowledge as we delve deeper into the mysteries of fractals.

REFERENCES

1. Barnsley, M. F., and Demko, S., Iterated function systems and the global construction of fractals, *Proceedings of the Royal Society of London. A. Mathematical and Physical Sciences*, Vol. 399(1817), pp. 243–275 (1985).
2. Ezhumalai, A., Ganesan, N., and Balasubramaniyan, S., An extensive survey on fractal structures using iterated function system in patch antennas, *International Journal of Communication Systems*, Vol. 34(15), p. e4932 (2021).
3. Yu, Z. G., Anh, V., and Lau, K. S., Iterated function system and multifractal analysis of biological sequences, *International Journal of Modern Physics* B, Vol. 17(22n24), pp. 4367–4375 (2003).
4. Singh, S. L., Prasad, B., and Kumar, A., Fractals via iterated functions and multifunctions, *Chaos, Solitons & Fractals*, Vol. 39(3), pp. 1224–1231 (2009).
5. Garg, A., Negi, A., Agrawal, A., and Latwal, B., Geometric modelling of complex objects using iterated function system, *International Journal of Scientific & Technology Research*, Vol. 3(6), p. 6 (2014).
6. Hart, J. C., Fractal image compression and recurrent iterated function systems, *IEEE Computer Graphics and Applications*, Vol. 16(4), pp. 25–33 (1996).
7. Kominek, J., Advances in fractal compression for multimedia applications, *Multimedia Systems*, Vol. 5(4), pp. 255–270 (1997).
8. Vrscay, E. R., Iterated function systems: Theory, applications and the inverse problem, In: J. Bélair and S. Dubuc (eds.), *Fractal Geometry and Analysis*, Springer, Netherlands, Dordrecht, pp. 405–468 (1991).

5 Fractals from Iterated Function Systems

5.1 INTRODUCTION

Fractals exhibit the property of approximate self-similarity, which is not perfectly accurate. These fractals are considered complex patterns at infinitely small scales. The parts of a self-similar object resemble each other. This approximate self-similarity occurs at progressively smaller scales. It describes many irregularly shaped or partially non-uniform objects in nature, such as mountain ranges, clouds, coastlines, lightning bolts, snowflakes, and more. This phenomenon is referred to as nominal self-similarity. The design is similar at various scales, but only approximately or not accurately. For example, the Mandelbrot set and Julia sets are examples of approximately self-similar fractal graphs [1].

This type of fractal appears similar at different levels of magnification. Fractals generated from iterated function systems (IFSs) stimulate artistic creativity and aesthetic exploration in addition to their mathematical and practical applications. By harnessing the creative potential of IFS algorithms, artists explore themes of symmetry, chaos, and recursion in their captivating visual compositions. The fractal beauty of IFS-generated images invites reflection on the relationship between pattern and randomness, order and chaos, and serves as a canvas for artistic expression. Finally, fractals from IFSs provide a synthesis of artistic inventiveness, computational efficiency, and mathematical elegance. As we delve deeper into the world of IFS fractals, we discover the beauty of self-similarity, the attraction of endless recursion, and the boundless opportunities for creative inquiry.

Fractals from IFSs are valuable in many different fields, such as scientific modeling, data compression, and computer graphics. IFS algorithms in computer graphics facilitate the creation of visually stunning visuals that replicate natural events with an unmatched level of realism. Additionally, IFS-based methods are employed for data compression, leveraging the self-similarity inherent in fractal patterns to achieve efficient digital information storage and transfer. Fractals from IFSs are also instrumental in scientific modeling, as they illuminate the complexity of natural systems and enable the exploration of phenomena such as diffusion, turbulence, and biological growth [2].

5.2 FRACTALS IN ALGEBRA

Fractals can form when a basic calculation is repeated multiple times, with the result fed back into the original. Benoit Mandelbrot, who coined the term "fractal" in the 1960s to describe the concept of fragmentation at all scales, is honored by the name of the first fractal we examine. Fractals are remarkable formations known for their

DOI: 10.1201/9781003481096-5

limitless complexity and self-similarity. They are widely used in various sectors and can be observed in nature, art, and computer-generated imagery. This essay explores the iterative algorithms that create fractals and their mathematical foundations. The mysterious shapes known as fractals, which emerged from the depths of mathematical study, have long fascinated intellectuals, scientists, and artists.

These geometric marvels were first introduced to the public by mathematician Benoit B. Mandelbrot in the 1970s, profoundly changing our understanding of complexity, shape, and space. The term "fractal" is derived from the Latin word "fractus," meaning shattered, due to a unique characteristic known as self-similarity. Each component of a fractal resembles the whole when viewed at different scales—an amazing property that allows for infinite levels of detail. Fractal geometry is defined by its self-referential nature, which underpins its incredible complexity.

5.2.1 Mathematical Foundation of Fractals

Understanding self-similarity is essential to comprehending fractals. When a fractal pattern is self-similar, it signifies that a portion of it is a smaller version of the entire structure. In other words, the pattern is repeated at various scales. This recursive feature is what contributes to the limitless complexity observed in fractals. In mathematics, recursive methods and IFSs are used to express self-similarity. The complex patterns of the fractal evolve through a series of affine transformations used in IFSs, with each transformation building on the preceding iteration to produce the next level of detail. Fractal generating software is employed to create images of fractals. These images are typically referred to as fractals, even when they exhibit the aforementioned characteristics. It is quite possible to zoom in on regions that display any fractal properties.

5.3 IFSs

A large number of classic fractals are generated via IFSs. Each linear combination of affine transformations, such as translations, rotations, and scaling, defines a contractive transformation, which comprises an IFS. The beauty of an IFS lies in the fractal structure that emerges when these transformations are applied repeatedly to a starting point, demonstrating self-similarity over several iterations. Hausdorff dimension: Traditional Euclidean geometric objects categorize points as zero-dimensional, lines as one-dimensional, and planes as two-dimensional. However, the fractional dimensions of fractals challenge this notion. The "dimension" of fractals is defined by the Hausdorff dimension, which indicates that fractals possess non-integer or fractional dimensions as opposed to integer dimensions.

For example, the Sierpinski triangle has a dimension $\frac{\log 3}{\log 2} = 1.58$, while the Koch snowflake has a dimension $\frac{\log 4}{\log 3} = 1.26$

Fractal Dimension: It generalizes the concept of dimension to include fractional values, in contrast to the conventional Euclidean dimension, which only deals with whole integers. A line, for instance, has a Euclidean dimension of 1, but a more complex fractal curve, such as the Koch snowflake, has a fractal dimension greater than 1. As a fractal's detail increases with more iterations, its fractal dimension approaches the value that best captures its intricate nature.

FIGURE 5.1 Mandelbrot set: classical iterated function systems are used to express self-similar images.

5.3.1 Mandelbrot Set

This type of fractal was examined by mathematicians Pierre Fatou and Gaston Julia in the 20th century and originated in complex dynamics. In 1980, Mandelbrot introduced the visualization of this set at IBM in New York. One of the most well-known fractals, the Mandelbrot set, explores the domain of complex numbers. A simple mathematical process involving complex numbers is iterated to define the Mandelbrot set. The characteristic form of the Mandelbrot set is revealed by analyzing whether the output of these iterations tends toward infinity or remains confined within a certain region. Figure 5.1 shows an image of the Mandelbrot set.

5.3.2 Julia Set

The concept involves quasi self-similarity in fractals, where similar parts are repeated throughout the whole. This is achieved through an iterative method. Another family of fractals closely related to the Mandelbrot set is termed Julia sets. Every point in the complex plane represents a distinct Julia set, displaying patterns that are both singular and captivating. By investigating complex iteration dynamics at various initial locations, Julia sets present a wide range of fractal configurations. The Julia sets and the Mandelbrot set are connected through fractal basins. Based on the convergence behavior of complex iterations, they partition the complex plane into areas, revealing the intricate boundaries that define fractals.

The Julia set is commonly denoted as Ju(f), while the Fatou set is denoted as Fa(f). In the early 20th century, Julia and Pierre Fatou conducted research on these sets, which is why they bear their names. A fractal is characterized by its self-similarity and symmetry across multiple scales. Julia set fractals represent a specific type of fractal defined by the behavior of functions that operate on complex numbers. These fractals depend on complex numbers, which consist of both real and imaginary components. To create the mathematical function f(z), which relies on complex numbers, the Julia set is described by the equation $z = z^2 + c$.

Figure 5.2 presents an example of a Julia set.

5.4 STATISTICAL SELF-SIMILAR FRACTAL GRAPHS

A branch of a tree or a frond from a fern exemplifies self-similarity [3]. Trees and ferns are natural fractals that can be modeled on a computer using a recurrence algorithm. This recursive nature is evident in examples such as a tree branch or a fern frond, which serve as miniature scaled-down imitations of the whole structure. They are similar in nature but not identical. Random fractals are considered statistical self-similar fractals, whether uniform or otherwise, and exhibit self-similarity in an approximate pattern across various scales. For example, the coastline paradox shows that the length of a coastline, which has a non-regular shape, can be infinite. Figure 5.3 illustrates the coastline paradox.

FIGURE 5.2 Julia Set.

FIGURE 5.3 Coastline paradox.

5.5 OVERVIEW OF FRACTAL ITERATED FUNCTION SYSTEMS

Fractals are created using mathematical models called IFSs. They consist of a collection of functions that are applied repeatedly to sets or points in space. A function set (F) is composed of several functions, most of which are affine transformations such as shears, translations, rotations, and scalings.

Probabilities (P): A probability weight is assigned to each function in the function set. These weights determine the likelihood of choosing a specific function during the iteration phase. The initial set, or S, is where the iteration begins. It can be any geometric shape, a single point, or a segment of a line. IFSs create self-similar structures by repeatedly applying transformations. Finer details of the fractal emerge as the iterations progress, displaying self-similarity at various sizes [4].

Computational Illustration: Mathematically, fractals produced by IFSs can be expressed as the fixed points of a collection of contraction mappings. The set that the iterations converge towards, signifying the final fractal shape, is known as the attractor of the IFS. Examples include the Koch snowflake, Barnsley fern, and Sierpinski triangle, which are well-known fractals generated by IFSs. Each fractal displays distinct geometric characteristics and self-similar patterns that arise from the underlying transformations.

IFSs provide a strong foundation for constructing and researching fractals, shedding light on the intricate details of mathematical and natural phenomena.

5.6 ITERATIVE PROCESS IN GENERATING FRACTALS USING IFS

There are various processes involved in the iterative creation of fractals with IFSs:

5.6.1 Starting Point

Begin with a collection of points or forms in space. This could be any geometric figure, such as a triangle, a single point, or a line segment. Repetition: Select a function from the function set, with probabilities determined by corresponding weights. Apply the chosen function to all points in the current set. This involves transforming each point according to the selected function. Continue this process until a termination condition is met or for a predetermined number of iterations. Revision Set: The points that remain after the function has been applied to each point in the current set create a new set. This updated set serves as the input for the next iteration. Randomly select functions from the function set and apply them to each point in the current set to repeat the operation iteratively. Continue iterating until you achieve the desired level of detail or reach a fractal attractor.

5.6.2 Convergence

As the iterations progress, the set of points converges on a fractal attractor.

The final form or pattern that emerges from the iterative process is known as the fractal attractor, representing the self-similar structure generated by the IFS. To visualize the fractal attractor, employ appropriate rendering methods.

This may involve creating a digital representation of the fractal, generating an image, or plotting the points in 2D or 3D space, depending on the application. IFS produces complex fractal designs characterized by fine detail and self-similarity at all scales. The diversity introduced into the fractal formation process by the probabilistic selection of functions results in a variety of aesthetically pleasing and mathematically complex structures.

5.7 INTEGRATION OF IFSs WITH OTHER MATHEMATICAL AND COMPUTATIONAL TECHNIQUES

There are intriguing opportunities to enhance fractal production, analysis, and use through the integration of IFSs with other mathematical and computational tools. The following are some ways IFSs can be combined with these techniques:

5.7.1 Estimating Fractal Dimensions

Integrate IFSs with fractal dimension estimation techniques, including Hausdorff dimension algorithms, correlation dimension, and box counting. This integration allows for a more accurate quantification of the complexity and self-similarity of the fractal patterns produced by IFSs [5].

5.7.2 Optimization Techniques

Apply optimization techniques to determine the best combinations of functions and probabilities in an IFS that yield the desired fractal designs or attributes. Algorithms such as particle swarm optimization, simulated annealing, and genetic algorithms can be employed to find the parameters that best match a given target fractal.

Wavelets Analysis: Use IFSs with wavelet analysis techniques to decompose fractal signals or images into distinct scales and frequencies. Fractals can be represented in several resolutions using wavelet analysis, which can reveal hidden features and structures at different levels of detail.

5.7.3 Artificial Intelligence

Utilize the fractals produced by IFSs to apply machine learning algorithms for tasks involving pattern recognition, classification, or generation.

Train neural networks or other machine learning models using datasets of IFS-generated fractals to discover patterns and relationships that facilitate tasks such as image recognition and synthesis.

5.7.4 Dynamic Systems

Combine IFSs with dynamical systems theory to investigate how fractals behave and change over time. Examine the stability, bifurcations, and attractors of fractals created using IFSs under various dynamical regimes to gain insights into their long-term dynamics and behavior.

5.7.5 Analytical Statistics

Analyze fractal features produced by IFSs using statistical techniques, such as multifractal spectrum, geographic distribution, and correlation structure.

Statistical methods, including Markov chain modeling, spatial autocorrelation, and fractal analysis, can help characterize the statistical characteristics of fractals formed by IFSs.

5.7.6 Visual Aids and Graphic Design

Utilize advanced rendering and visualization techniques to visualize and portray the intricate fractal patterns produced by IFSs.

For scientific visualization, art, or entertainment, IFS-generated fractals can be rendered into high-quality visualizations using methods such as procedural generation, volumetric rendering, or ray tracing. Researchers can gain fresh insights into the vast diversity of fractal patterns across various fields by combining IFS with these and other mathematical and computational methods. This multidisciplinary approach creates exciting potential for innovation and discovery while fostering collaboration across multiple disciplines.

5.8 IMPORTANCE OF IFSs IN THE STUDY AND APPLICATION OF FRACTALS

IFSs are valuable for studying and applying fractals due to their mathematical rigor, adaptability, and ability to generate intricate and aesthetically pleasing patterns. The IFS is important in this field for the following reasons:

5.8.1 Generative Power

IFSs provide a structured and efficient method for producing a wide range of fractal patterns and shapes. Rich geometric features are produced by IFSs, creating complex and self-similar structures. IFSs provide an accurate representation of fractal geometry, facilitating mathematical analysis and modeling.

5.8.2 Computational Efficiency

Because IFS algorithms are computationally efficient, they enable effective generation and manipulation of fractals in real-time applications. This efficiency allows for the rapid exploration and visualization of fractal patterns in areas such as virtual reality, simulation, and computer graphics.

5.8.3 Self-Similarity

Fractals produced by IFSs exhibit self-similarity, a property characterized by similar patterns at different scales. This feature is essential for simulating fractal-like natural processes and phenomena, such as clouds, biological forms, and coastlines.

Versatility: IFSs can generate fractals with varying levels of complexity and visual appeal. By manipulating the function set and probability factors, researchers can create a diverse range of fractal forms, from simple geometric designs to intricate natural formations.

5.8.4 Multidisciplinary Applications

IFS is studied and applied across various fields, including computer science, physics, art, engineering, and mathematics. Applications for IFS-generated fractals encompass texture synthesis, generative art, data visualization, and image compression.

5.8.5 Educational Value

IFS serves as an educational tool that helps students understand fractals and the fundamental concepts of recursion and self-similarity. Through hands-on experiences in creating and analyzing fractal patterns, students can develop a deeper understanding of complex systems and geometric concepts by exploring IFS algorithms.

In summary, IFSs are essential to the study and application of fractals due to their generative power, mathematical formalism, computational efficiency, and interdisciplinary relevance. By utilizing IFSs, scholars and professionals can explore the vast variety of fractal geometry and gain new insights into the intricate patterns observed in both natural and mathematical domains.

5.9 FUTURE PROSPECTS AND POTENTIAL IMPACT OF FRACTALS FROM IFSs ON SCIENCE AND TECHNOLOGY

Fractals produced by IFSs have a wide-ranging and exciting prospective impact on science and technology in the future. Fractals from IFSs could significantly influence the following areas:

5.9.1 Data Encoding and Compression

IFS-generated fractals have previously demonstrated potential in image encoding and compression techniques. Future developments in fractal-based compression methods may enable multimedia data to be stored and transmitted more effectively, impacting industries such as satellite imaging, medical imaging, and telecommunications.

5.9.2 Computer Visualization and Graphics

IFS fractals provide abundant opportunities for creating visually stunning simulations and visualizations. Advancements in rendering techniques and hardware acceleration could facilitate realistic special effects, interactive visualizations for science and entertainment, and immersive virtual environments.

5.9.3 Machine Learning and Pattern Recognition

This could benefit various industries, including banking, security, and healthcare.

5.9.4 Imaging and Analysis in Biomedicine

The use of fractal analysis to describe intricate biological processes and structures has shown promise. Fractals produced by IFSs may provide fresh perspectives on the fractal nature of biological systems at different scales, advancing biomedical imaging methods, disease diagnostics, and drug discovery.

5.9.5 Engineering and Science of Materials

Applications for IFS fractals include modeling surface roughness, fractal aggregates, and porous materials. Future studies could lead to the development of unique materials with specific features for use in catalysis, energy storage, filtration, and nanotechnology by utilizing fractal-based simulations and analysis.

5.9.6 Environmental Tracking and Evaluation

Environmental phenomena such as river networks, weather patterns, and biological systems have been studied using fractal analysis [1]. Fractals produced by IFSs may aid in the creation of predictive models, risk assessments, and mitigation plans for environmental issues, including climate change, natural disasters, and ecosystem management.

5.10 CONCLUSION

In conclusion, fractals produced by IFSs have the potential to significantly advance science and technology across various domains, including computer graphics, biological imaging, and environmental monitoring, in addition to data compression. Fractal geometry and its applications present opportunities for solving complex problems and developing innovations that will benefit society, provided further research and creativity are invested in them.

REFERENCES

1. Riehl, J. R., and Hespanha, J. P., Fractal graph optimization algorithms, In *Proceedings of the 44th Industrial of Electrical and Electronics Engineers Conference on Decision and Control*, Seville, Spain, pp. 2188–2193 (2005).
2. Kumar, R., and Kant, R., Graph theory, importance and scope, *International Global Journal for Research Analysis*, Vol. 4(7), pp. 365–367 (2015).
3. Ore, O., *Theory of Graphs*, Vol. 38, American Mathematical Society Colloquium Publications, Providence, RI (1962).
4. Lai, C., On the size of graphs with all cycle having distinct length, *Discrete Mathematics*, Vol. 122(3), pp. 363–364 (1993).
5. Krön, B., Growth of self-similar graphs, *Journal of Graph Theory*, Vol. 45(3), pp. 224–239 (2004).

6 Application of Fractals in Various Fields

6.1 INTRODUCTION

Fractals have many applications in our everyday lives. Some of these real-life applications can be listed as follows: chaos is actually very structured and adheres to specific patterns. The challenge lies in locating these complex and elusive patterns. One goal of using fractals to explore chaos is to forecast patterns in dynamical systems that appear unpredictable at first glance. For many chaologists, the study of fractals and chaos represents a revolution rather than merely a new branch of research that combines computer technology, theoretical physics, art, and mathematics. It is the discovery of a new geometry—one that characterizes the infinite universe in which we exist; a universe that is ever-moving, unlike the still pictures seen in textbooks.

6.2 FRACTALS IN NATURE

The main uses of fractal geometry include the notion that objects created by God can be considered approximate fractals to a certain degree. This encompasses clouds, mountain ranges, coastlines, vegetables, and more. Thus, it provides a baseline for understanding the continuous patterns found in nature. Algorithms are developed using concepts from fractal geometry to simulate and comprehend objects in nature. Fractals have found applications in diverse fields. They are listed as follows:

- Modeling natural structures, geographical contours, and biological plant structures
- Image compression in computer graphics
- Analyzing diagnostic images in medicine
- Applications in architectural engineering
- Studying the convergence of iterative processes
- Investigating chaos phenomena
- Artistic endeavors

6.3 FRACTALS IN CELLULAR SYSTEMS

Antenna design using fractals leads to clear and efficient designs. Traditional antenna designs are often sensitive and provide a narrow range of frequencies. They become inefficient if they are smaller than a quarter of the wavelength, which is a problem for small, portable antennas such as those found in cellular phones. However, fractal antenna designs overcome these challenges. Experiments have shown that antennas built with a minimal number of iterations of the fractal process can demonstrate

DOI: 10.1201/9781003481096-6

FIGURE 6.1 Fractal design in mobile phones.

sensitivity across different frequencies. As the number of iterations increases, the antenna operates at lower frequencies while also accommodating higher frequencies. Additionally, fractal antennas are effective at one-quarter the size of traditional designs. Properly harnessed, these features represent significant advantages. Several companies are using fractals to create compact and multi-frequency antennas in their cellular phones, and military communications hardware exhibits similar characteristics.

A company focused on fractal antennas was founded by Nathan Cohen, who designed a cellular phone antenna based on the Sierpinski Carpet design. This antenna is integrated within the body of the phone. The multi-frequency capability of the antenna allows GPS to be incorporated into the device, as shown in Figure 6.1. Other applications include multi-frequency wireless LANs and maritime antennas.

6.4 FRACTALS IN COMPUTER GRAPHICS

The elements of fractal geometry and its applications are utilized in computer graphics and geometric modeling. A connection with chaos dynamics, primarily from a historical perspective, is emphasized. Two fundamental algorithms for computing fractal attractors are described. Iterated function systems (IFSs) are presented as a means of constructing deterministic fractals. The introduction of parameters in IFSs, through Bernstein polynomials, is discussed to produce various natural forms. A variety of applications in animation, data compression, rendering objects, and modeling phenomena in physics and biology are highlighted.

6.5 ASTRONOMY

Fractals may be transforming our understanding of the universe. Most cosmologists believe that matter is dispersed evenly throughout space. At small scales, astronomers generally support this premise; however, at very large scales, the majority contend that the cosmos is smooth. In contrast, a minority of scientists argue that the universe's structure is fractal across all scales. Consequently, galactic structures exhibit extreme asymmetry and self-similarity [1]. However, more information about the distribution of matter in the universe is currently needed by cosmologists to confirm—or refute—the theory that our universe is fractal.

6.6 COMPUTER SCIENCE

Fractal image compression is the most practical application of fractals in computer science. This type of compression leverages the fact that fractal geometry provides an excellent description of the real world. This method compresses images significantly more than standard techniques (such as JPEG or GIF formats). Another benefit of fractal compression is the absence of pixelation when an image is enlarged; increasing the size of the picture often makes it appear clearer.

6.7 FLUID MECHANICS

The study of flow turbulence is well-suited to the examination of fractals. Turbulent flows are unpredictable and challenging to model accurately. To aid engineers and physicists in better understanding complex flows, these flows can be represented as fractals. Flames can also be mimicked in this manner. Fractals provide an effective way to visualize porous materials due to their intricate geometric structures. In fact, petroleum science utilizes this concept. One of the most prevalent consequences of diabetes in ocular diseases is diabetic retinopathy. Diabetes damages the retina's blood vessels, leading to this condition, which is one of the main causes of blindness in diabetics.

6.8 FRACTALS AND DIABETIC RETINOPATHY

After pupil dilation, an ophthalmologist examines the fundus of the eye as part of the standard and traditional method for screening diabetic retinopathy. This type of assessment varies based on the treating physician's opinion, and subjectivity still exists, particularly in the early stages of the condition. In recent decades, novel approaches to detecting diabetic retinopathy have been explored. The primary instrument used in fractal studies is the fractal dimension, one of the metrics for describing blood network complexity [2]. Numerous investigations into the formation mechanism of the retina's vascular network have produced outstanding results, leading to the development of new interpretations based on the assessment of the fractal dimension using various algorithms or methods [3]. This is illustrated in Figure 6.2, while Figure 6.3 shows examples of diabetic retinopathy.

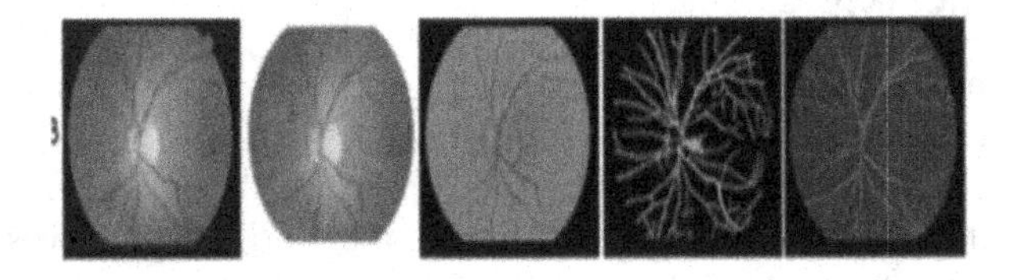

FIGURE 6.2 Image of normal retinal vessel network.

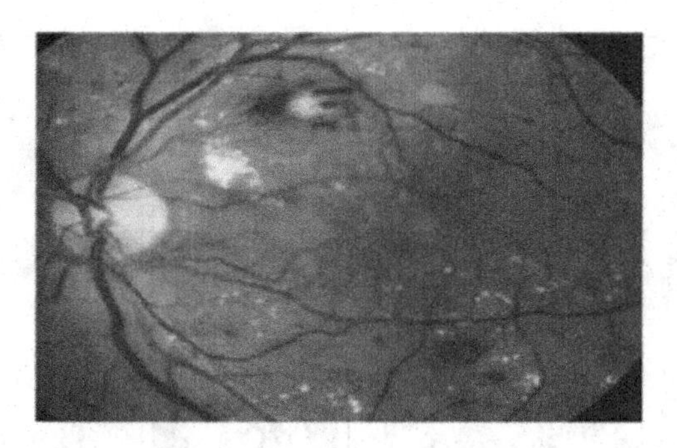

FIGURE 6.3 Diabetic retinopathy.

6.9 CANCER TISSUES

A prerequisite for the fractal approach's objective description of pathogenic and neoplastic features in cell A prerequisite for the fractal approach's objective description of pathogenic and neoplastic features in cell tissues is the experimental definition of a scaling interval, as opposed to a distinct dimensional scale chosen in advance. Several examples suggest that constructive mechanisms, associated with fractal principles—such as deterministic and/or random iteration of constituent units with varying degrees of self-similarity, scaling properties, and form conservation—may be responsible for the occurrence of morphogenetic dynamics, the emergence of complex patterns, and the architectural organization of active tissues and tumor masses. Clinicians trained to categorize aberrant traits, such as structural abnormalities or high mitotic indexes, usually interpret these observations qualitatively. Herein lies the potential of fractal analysis as a morphometric measure of the irregular structures typical of tumor growth. Figure 6.4 shows the mean box fractal dimension and the number of cases for different age groups.

This study demonstrates how fractal graphs are applied in many fields, particularly in medicine [4]. The fractal dimension of brain tumors differs from that of other organ sections. The expansion of blood vessels in other organ sections causes variances in cell dimensions. We believe that fractal geometry [1] provides insights into the morphology of tumors and evolves into a useful tool for analyzing the complicated and irregular structures of human body cells. This essay examines diabetic retinopathy, one of the most prevalent complications of diabetes. The vascular network of the retina forms through a mechanism based on the assessment of the fractal dimension using various methods.

6.10 SIGNATURE RECOGNITION BY FRACTAL DIMENSIONS

A person's signature is a handwritten, stylized version of their name. The definition of a "signature" derives from the words "sign" and "nature," meaning that your signature reflects your nature. A signature is not the result of a virtual activity and

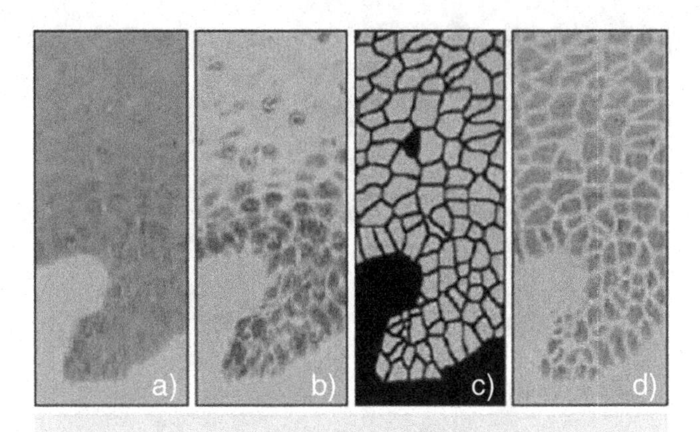

Mean box fractal dimension and the number of cases for different age groups

Age range	Mean box fractal dimension	Number of cases
0-10	1.1069 ± 0.03267	2
11-20	1.1117 ± 0.08987	2
21-30	1.1414 ± 0.07095	4
31-40	1.1290 ± 0.03822	4
41-50	1.0933 ± 0.05661	5
51-60	1.1373 ± 0.05657	8
61-70	1.0992 ± 0.04456	9
71-80	1.1260 ± 0.04751	5
81-90	1.0903 ± 0.01675	3

FIGURE 6.4 Characteristics of different persons' signature and fingerprints.

cannot depend on fractal geometry alone [5]. The characteristics of the fractal phenomenon must be demonstrated, and various signatures must be analyzed. A person's signature is a unique graphic representation of their personality, defined both consciously and subconsciously. It is not merely handwriting; it is a freehand sketch of oneself, allowing the individual to breach conventional standards and norms of handwriting without question. Whether the signature is opaque, somewhat opaque, or complex is entirely up to the individual. Nonetheless, there should be readability and transparency in the writing. Since the purpose of a signature is to establish a personal brand through a handmade image, handwriting serves a different purpose than simply signing. As a result, there is significantly more creative freedom when designing a signature compared to traditional writing. It is possible to describe the parameters surrounding signature writing and categorize the various styles that emerge. Because signatures differ across populations, they are fundamentally a human invention. To highlight the distinctions between fingerprints and signatures, it is important to first determine the features of fingerprints, including the mean box fractal dimension and the number of cases for different age groups.

Different fractal dimensions will undoubtedly arise from the signatures of different individuals (for both the D_B and D_D methods) [6]. While neighborhood influences may contribute to these differences, the writer's behavior and possibly even their character are the primary causes. Each individual is born with a distinct chromosomal program that differs from that of their parents, resulting from a combination of two parental programs. Once established, this program remains constant for the individual's lifetime. Graphologists eagerly examine the unique characteristics of signatures, searching for details that reveal a person's personality. Additionally, DNA fingerprinting may alter the permanent character mark present in signatures, which could be useful in the genetic diagnosis of cancer.

There are many methods for intuitively understanding the behavior and structure of fractals. Fractals are a class of mathematical structures used to model various types of time series data and images, characterized by great geometrical complexity. A key feature of fractals is their fractal dimension, which provides insight into their geometric configuration. The growing interest in fractal geometry is leading to an increasing number of applications in other fields. This section intends to demonstrate how similar concepts can also be applied to identify different people's signatures. Figures 6.5 and 6.6 shows the characteristics of signatures of different persons.

Using fractal dimension aims to develop a new technique for identifying distinct signatures of individuals [7]. Here, we investigate the composition of different signatures and the possible applications of several recently introduced techniques to extract more information. These methods include the utilization of fractal characteristics, which enhance DNA pattern discrimination, and principal component

FIGURE 6.5 Signatures of different persons.

FIGURE 6.6 Analysis of signatures from various persons using the box counting method.

analysis, which characterizes the morphology of an area. Numerous diseases have been diagnosed using these techniques, yielding incredibly positive results. Fractals are a useful tool for studying complexity. DNA pattern recognition requires characteristics to be extracted from specific regions of an image and then processed with a Cantor set pattern classification.

One of the primary characteristics is size. The dimension of a fractal item can be used to interpret how much "space" it occupies between dimensional manifolds and random metrics. Form analysis, segmentation, texture classification, and other problems have been addressed using these properties. Here, we use two methods to estimate the fractal dimension of different people's signatures: box counting dimension and divider step dimension. In Section 10.2 we discuss methods for obtaining signatures, such as the divider step dimension method and the box counting method. Based on the dimensions of the signatures, Section 10.3 explains principal component analysis for a variety of signatures. Each individual has a distinct DNA pattern; Section 10.4 explores how these fingerprints may be used in the future to diagnose a range of illnesses, such as HIV, TB, and hypertension.

6.10.1 Fractal Dimension Computational Methods

Any measurement that varies according to changes in observation size is called a fractal set. In the fractal context, the evolution law follows a power law. The value of the power indicates the complexity of the set in relation to the metric being studied. When employing a fractal technique to analyze a phenomenon, a fractal dimension can be defined by considering the feature that needs to be evaluated and for which the evolution can be shown to follow a power law. A Method Adapted for Signature Verification: The challenge lies in determining how long a drawing should be if it contains multiple anomalies, varying in number, related to different individuals' signatures. The goal is to automatically and impartially analyze signatures.

We will examine a set of pixels, or images of signatures. Depending on how the data are gathered, there are two methods for automatically recognizing a signature: online and offline. Offline signature recognition is based on an additional document scan. Since all characteristics of a signature depend on spatial variables, it is relatively easy to counterfeit one by simply mimicking the writing style. As the pencil or pen writes, data are continuously collected using a digitalization table and a digital pencil. These data include the pressure applied, the tilt of the pen or pencil, and other variables. For the movements and writing style of the signature to be captured in the database, the signer must be present during the online signature data registration process.

6.10.2 Box Counting Dimension Method

One of the most commonly used dimensions is the box counting dimension. Its relative ease of mathematical computation and empirical estimation largely contribute to its popularity. Let $N_\delta(F)$ be the small number of sets of breadth at most δ that can cover F, and let F be any non-empty bounded subset of R^n.

Algorithm 6.1

Level 1: A normal mesh with a mesh size of r is created from the image (signature).

Level 2: Determine N(r) how many square boxes intersect the picture.

Level 3: The number N(r) depends on the scaling (grid) option selected.

Level 4: We count the corresponding number N(r) after repeating for a number of size values.

Level 5: By plotting log N(r) against log(1/r) we create the slope D.

The following level indicates the size or complexity (slope) of the fractal. A straight line is fitted to the plotted points in the diagram using the least squares method. The following linear regression equation was utilized to determine the fractal dimension:

$$\log N(r) = \log K + D_B \log(1/r) \tag{6.1}$$

where K denotes a constant and D_B represents the fractal dimension. This process has been applied to multiple signatures, producing the dimensions for each signature (Figure 6.7).

Figure 6.6 shows the analysis of signatures from various persons using the box-counting method

Figure 6.8 provides visual depiction of signatures from many individuals.

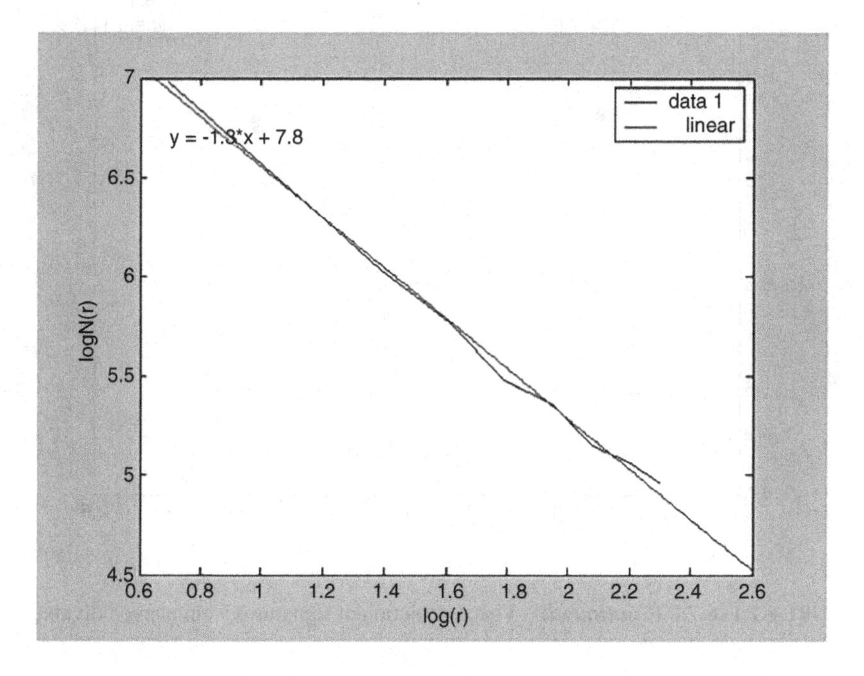

FIGURE 6.7.1–6.7.6 Visual depiction of signatures from many individuals.

(Continued)

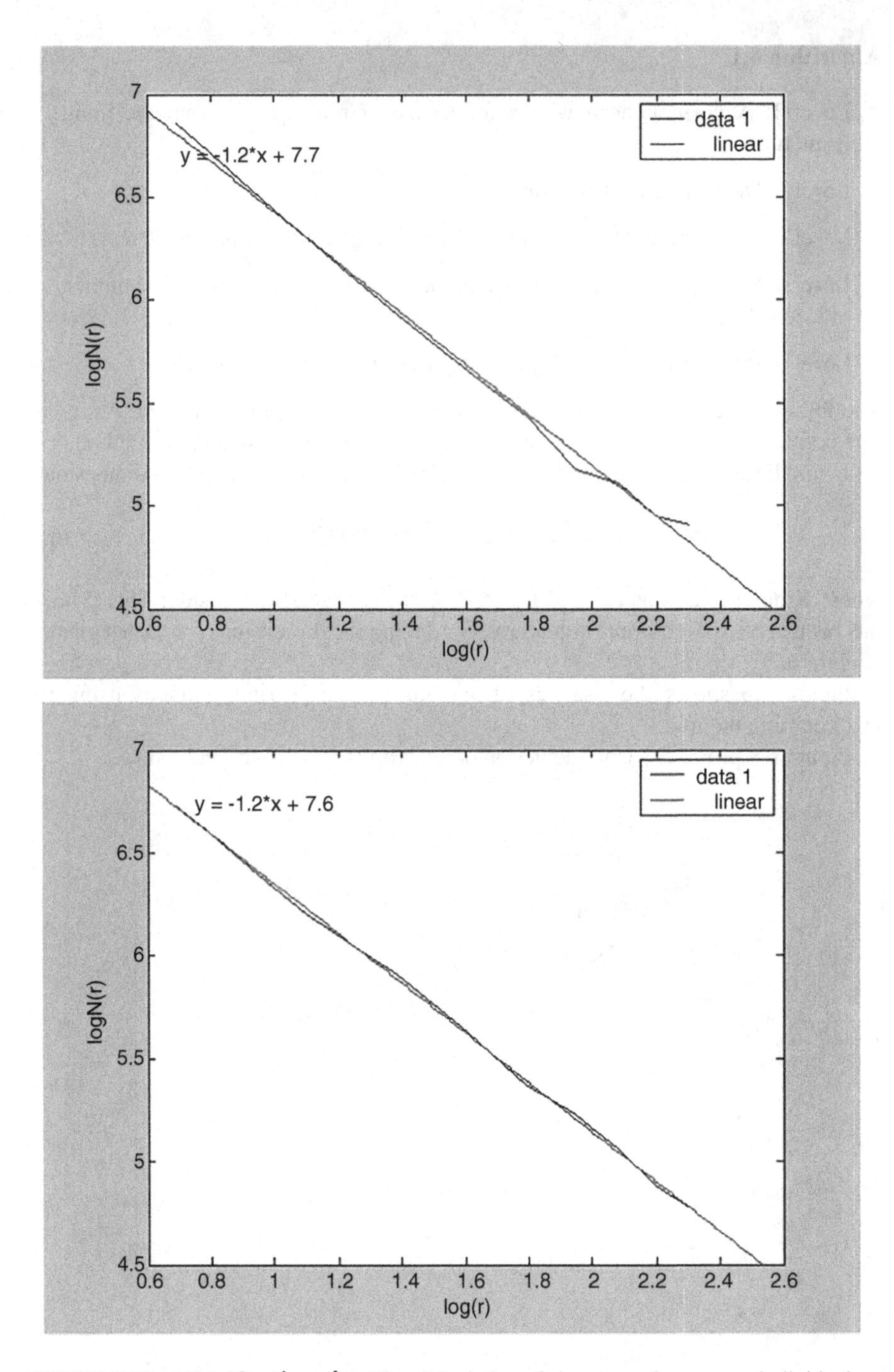

FIGURE 6.7.1–6.7.6 (*Continued*) Visual depiction of signatures from many individuals.

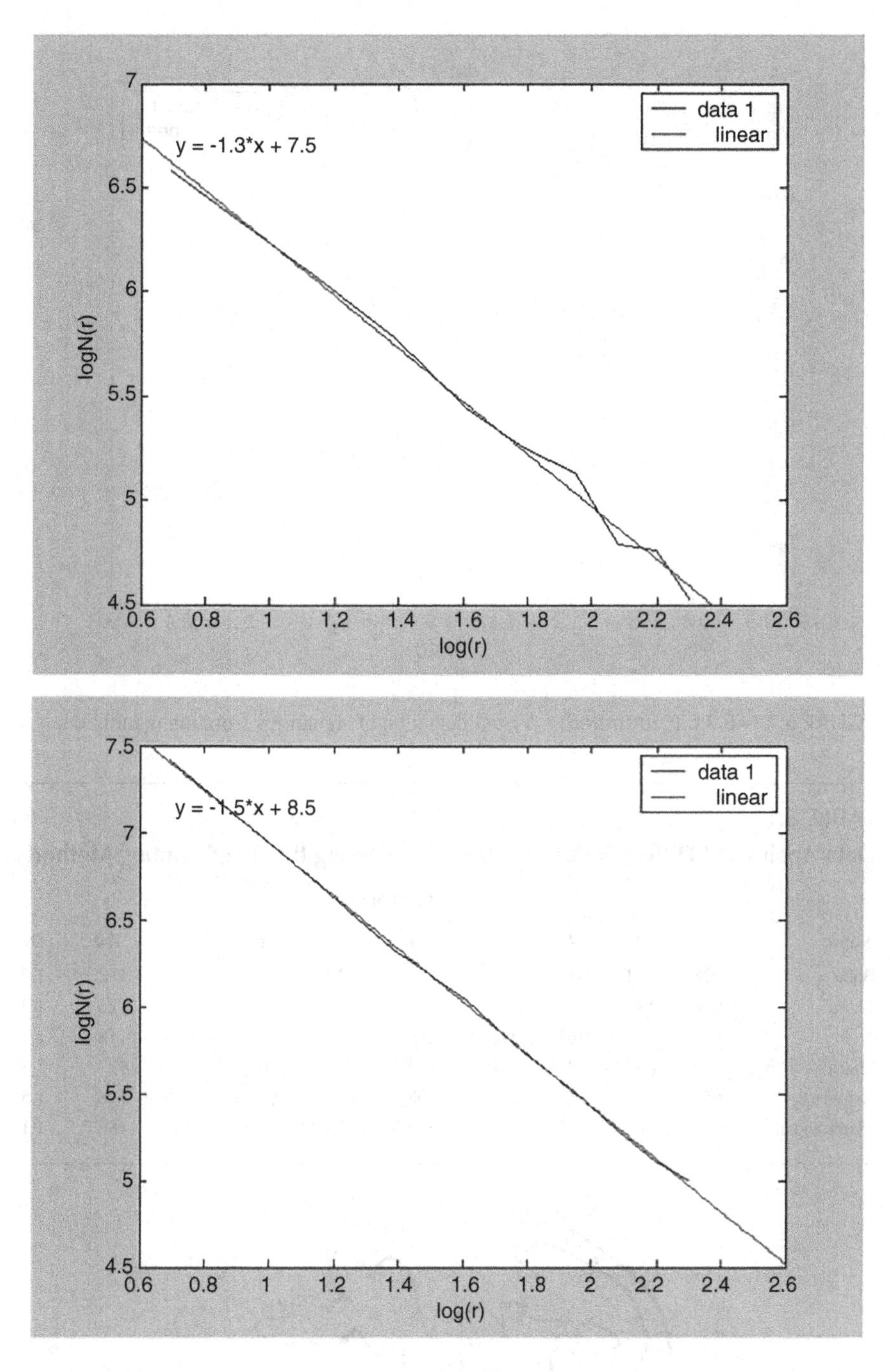

FIGURE 6.7.1–6.7.6 (*Continued*) Visual depiction of signatures from many individuals.

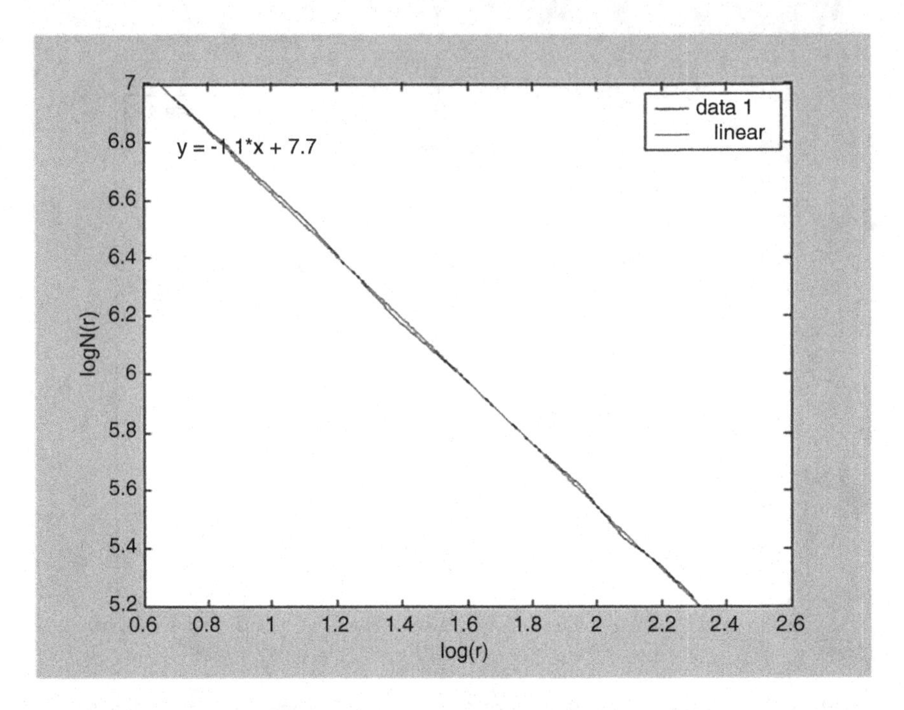

FIGURE 6.7.1–6.7.6 (*Continued*) Visual depiction of signatures from many individuals.

TABLE 6.1
Data Analysis of Different Persons' Signatures Using the Box Counting Method

	Grid Size									
Name	**2**	**3**	**4**	**5**	**6**	**7**	**8**	**9**	**10**	D_B
Babu	1,078	621	418	320	239	212	172	159	142	1.3
Thara	948	545	377	289	230	177	166	141	125	1.3
Yuva	835	491	367	276	216	186	158	132	118	1.2
Sharmi	722	455	324	231	190	169	120	117	92	1.3
Ashwin	1,659	887	563	418	308	248	197	166	148	1.5
Manikandan	1,046	687	487	389.2	320.5	275	229.88	209	186.7	1.1

FIGURE 6.8.1–6.8.7 Graphical representation of different persons' signatures.

(*Continued*)

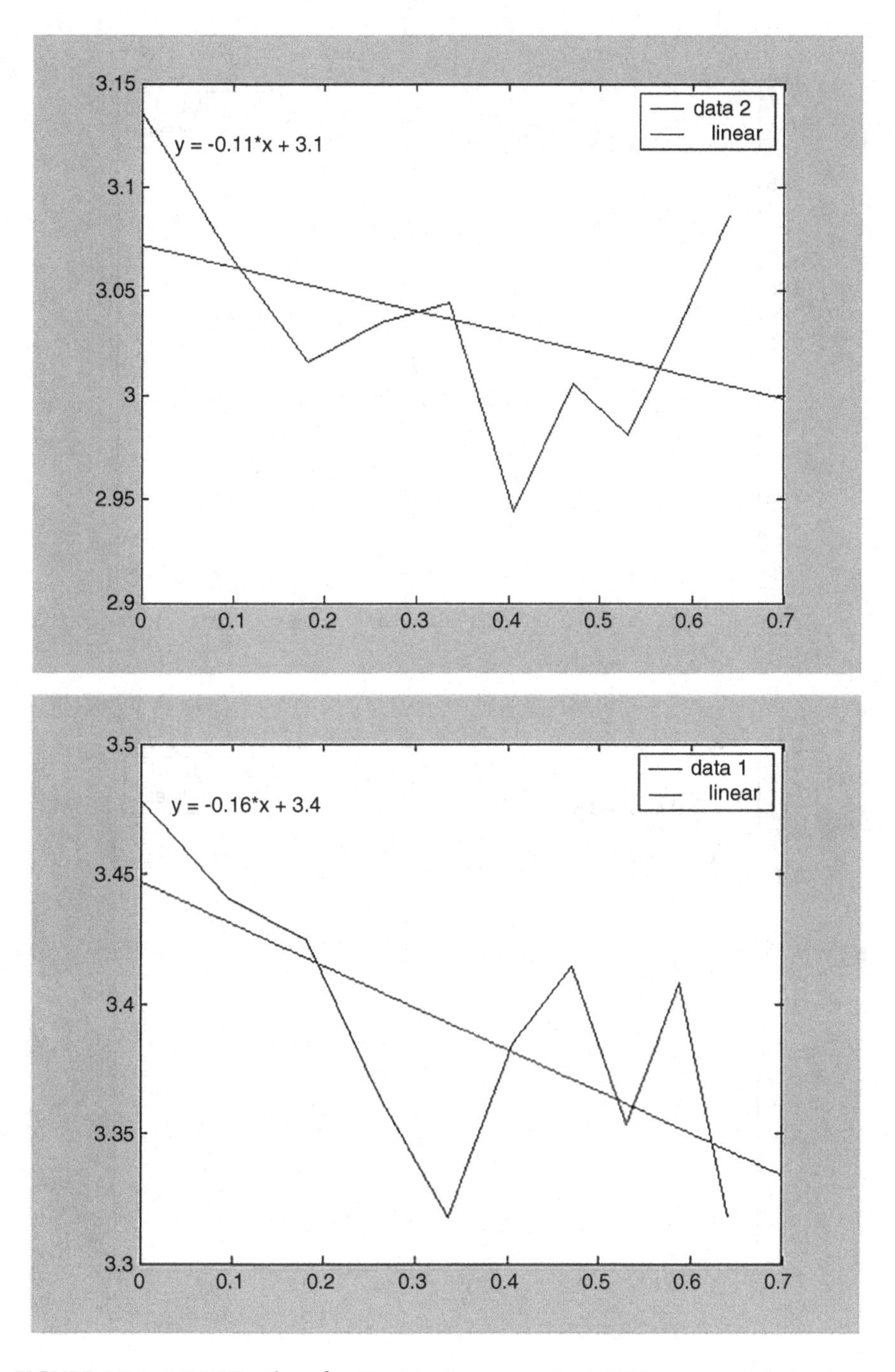

FIGURE 6.8.1–6.8.7 (*Continued*) Graphical representation of different persons' signatures.

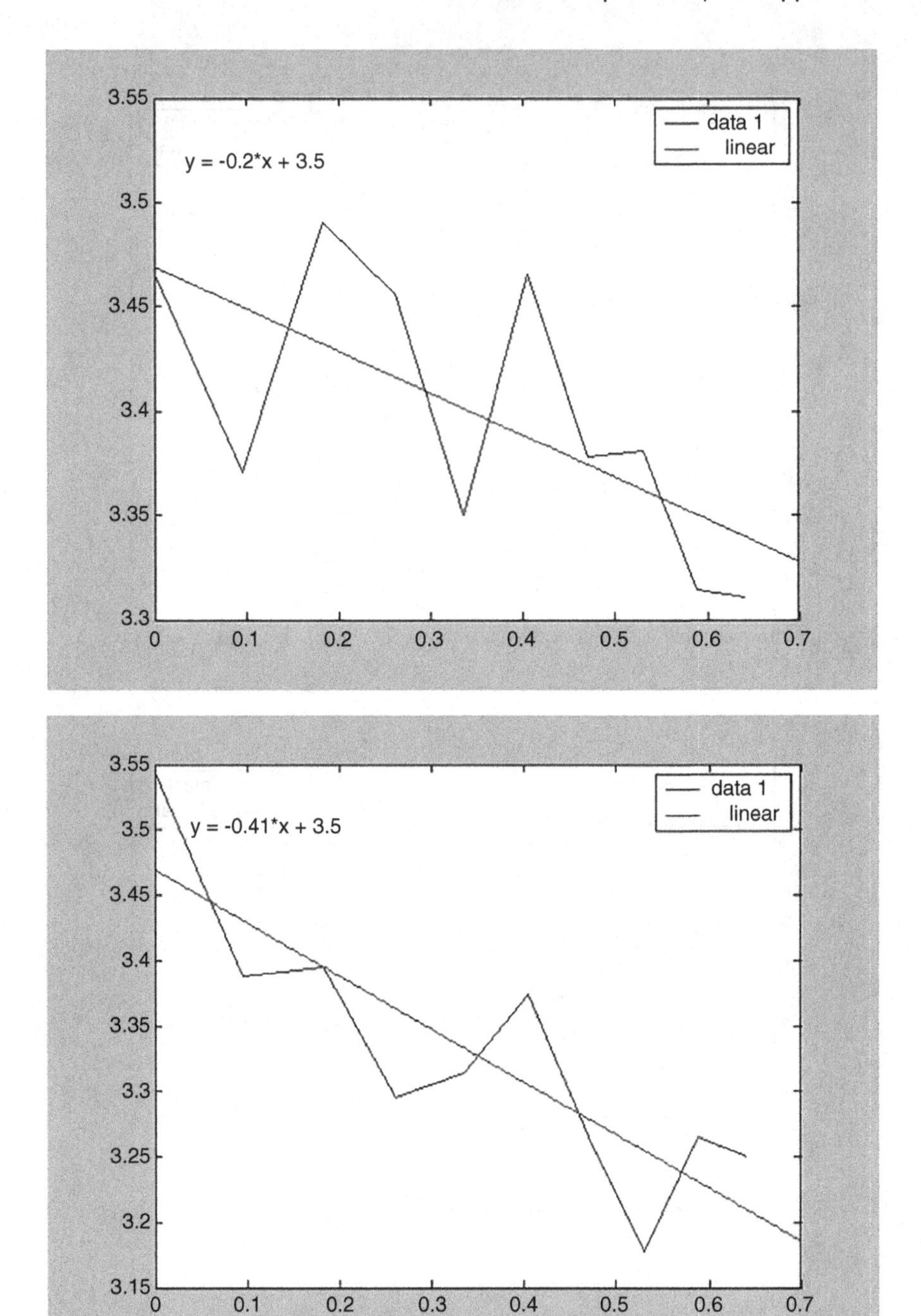

FIGURE 6.8.1–6.8.7 (*Continued*) Graphical representation of different persons' signatures.

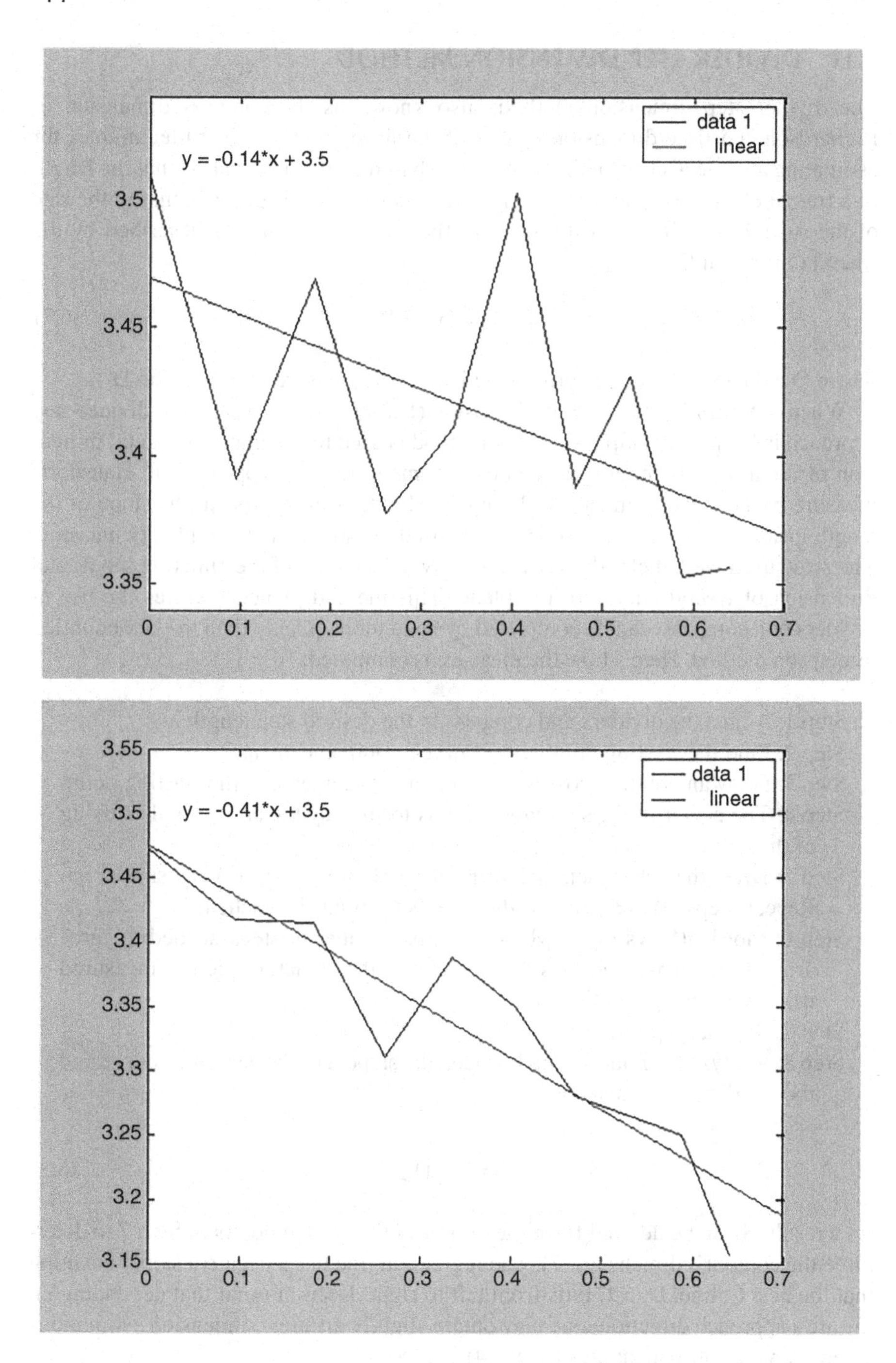

FIGURE 6.8.1–6.8.7 (*Continued*) Graphical representation of different persons' signatures.

6.11 DIVIDER STEP DIMENSION METHOD

The divider step dimension method, also known as the compass dimension or Richardson coastline dimension, is one of the most straightforward techniques for estimating an object's fractal dimension. It is based on the observation that the length of a fractal object's (signature image) contour in the plane is proportional to the size of the ruler λ, which is used to measure the contour's length, as described by the fractal dimension D.

$$L(\lambda) = D \cdot \lambda^{1-D} \tag{6.2}$$

where D belongs to R is an approximate Hausdorff measure in dimension D.

When determining the fractal dimension (D_D) of fractal objects, the divider-step approach is frequently employed. This method is used to calculate the fractal dimension of a curve or contour. Our complexity measure is an approximate Hausdorff measure in the fractal dimension. Using the divider-step approach, the slope of the length measured by various-sized rulers is used to estimate the complexity measure. The structured walk methodology is a widely used method for estimating the fractal dimension of fractal curves in the plane. This method, which requires the use of dividers or a compass, can be completed by hand more quickly than the box counting dimension method. Here's how the measure is computed:

Step 1: Adjust the dividers and compass to the desired step length λ
Step 2: Find the starting point at the curve's commencement.
Step 3: Draw an arc that crosses the curve and is centered at the starting point.
Step 4: The center of the subsequent arc is formed by the arc's initial crossing of the curve.
Step 5: Draw the subsequent arc with its center at the Step (4) crossing point. Repeat Steps (4) and (5) until the curve's terminus is reached.
Step 6: Plot log(L) vs log λ where N is the number of steps needed to "arc" around the curve, and L is the contour of the signature picture measured using λ as a step length. $L = N\lambda$
Step 7: Repeat Steps 1 to 6
Step 8: Analyze the relationship between the slope S of the generated curve and the divider step dimension.

$$S = 1 - D_D \tag{6.3}$$

As a result, S can be derived from the best fit of the plotted points in Step 7 to determine the contour's dimension (7). The regression line has a negative slope, meaning that the best fit line, $D_D > 1$, falls from left to right. Keep in mind that depending on the arc's approach direction, one may obtain slightly different dimension estimations when drawing subsequent arcs (Steps (4) and (5)).

Hence, the dimension of the contour may be determined by measuring S from the best-fit line of the plotted points in Step (7). The slope of the regression line is negative, that is, the best fit line falls from left to right, where $D_D > 1$. Note that when drawing successive arcs (Steps (4) and (5)) one may obtain slightly different dimension estimates depending upon the direction of approach of the arc. This can be programmed by MATLAB. Figure 6.3 provide visual depictions of signatures from various individuals, representing the graphical analysis of different persons' signatures and the examination of various signatories using the divider step dimension approach.

Connection between Fractal Parameters and the Character of the Signature: There is limited literature solely within computer science on this topic. For graphologists, it is also a significant and fascinating subject, offering important insights into various individuals. Graphologists have different objectives than we do and employ more traditional techniques. They believe that every human being is born with a prime genetic capital that determines their behavior, which cannot be altered, similar to our perspective. There is much to gain from comparing our results with theirs, providing an opportunity to validate our settings and refine their content. Additionally, graphology is used in medicine, particularly in hospitals, to help diagnose and monitor disorders of the nervous system and brain.

6.12 GRAPHOLOGIST PARAMETERS

The graphologist uses different parameters than we do. In our case, we are more concerned with sketching the writing, while the graphologist focuses primarily on the behavior of the signatory. They can obtain quantifiable findings by using the following parameters:

1. Graphical line's length (L). It is the length of the writing skeleton.
2. Velocity (V). This measures how long the signature lasts. The speed is determined by dividing the total writing time by the length of the signature skeleton.
3. Index of loop importance (B). This is calculated by determining the inner surface area of the loops.
4. Index of pressure (P). This is a significant and straightforward characteristic to measure in online writing.
5. Index of signature pressure (PS). This frequently differs from what is discussed in other sections of the book.

6.13 PRINCIPAL COMPONENT ANALYSIS

This method allows a set of variables to be transformed into a small number of linear composites that have the highest possible correlation with the original variables. For p provided variables, we can extract at most p principal components using this method. Occasionally, we encounter situations where the contribution ratio of the first principal component is lower than that of the subsequent components. Frequently, we find a principal component that lacks a significant interpretation corresponding to the second or third largest eigenvalue. Therefore, a technique is needed that allows us to extract fewer components using specific models (Tables 6.2 and 6.3).

TABLE 6.2
Principal Component Analysis of Box Counting Dimension for Different Signatures

Parameter	Babu	Tharaniya	Yuva	Sharmi	Ashwin	Manikandan
Length	7.8	7.7	7.6	7.5	8.5	7.7
Speed	1.81	2.56	4.50	4	7	7
Loop	51.685	46.463	47.778	40.574	63.778	69.690
White	3	4	6	13	12	15
Rapidity	1.7545	1.95	2.6809	4.0191	5.7273	5.2727
Juxta	1	1	1	2	2	3
Dim (D_B)	1.3	1.3	1.2	1.2650	1.5	1.1

Length = 99.3462%.
Velocity = 0.65603%.
Loop = 0.0706%.
White index = 0.0191%.
Quickness = 0.0037%.
Comparison = 0.0000%.

TABLE 6.3
Principal Component Analysis of Divider Step Dimension for Different Signatures

Parameter	Babu	Thara	Yuva	Sharmi	Ashwin	Manikandan
Length	3.1	3.6	3.6	3.5	3.6	3.6
Speed	1.81	2.56	4.5	4	7	7
Loop	7.8	7.7	7.6	7.5	8.5	7.7
White	3	4	6	13	12	15
Rapidity	1.7545	1.95	2.6809	4.0191	5.7273	5.2727
Juxta	1	1	1	2	2	3
Dim (D_D)	1.26	1.62	1.65	1.62	1.65	1.63

Length: 79.0088%.
Velocity = 19.4864%.
Loop = 1.1895%.
White index = 0.2452%.
Quickness = 0.0690%.
Comparison = 0.0011%.

6.14 FRACTAL MODEL FOR BLOOD FLOW IN THE CARDIOVASCULAR SYSTEM

Heart disease is becoming increasingly common in today's population. A heart attack occurs when the blood supply to a portion of the heart muscle is completely cut off. Understanding the anatomy and function of blood vessels is crucial for comprehending vascular disorders, particularly ischemia, which is the world's leading cause of mortality. Blood vessels are involved in nearly all medical disorders. The system consists of the heart, blood, and blood vessels, with the heart at its core. Blood vessels include arteries, veins, and capillaries. An aorta or other large artery divides into arterioles, which then branch into capillaries, the smallest blood vessels. Capillaries, in turn, create venules, which combine to form veins. The heart pumps blood through this vast network of vessels, serving a singular purpose. The hydrostatic pressure within the circulatory system varies, and the structure of the vessels adapts accordingly. In the immediate vicinity of the heart, where hydrostatic pressure is highest, blood vessels have the thickest and most complex walls. Figure 6.9 shows an image of the heart, and Figure 6.10 illustrates blood flow in the cardiovascular system.

Fractal processes, or time series, cannot be described by a single time scale, nor do fractal structures have a single length scale [8]. Here, we explore the cardiovascular system and present a potential fractal model of the blood vessel system. The "fractal model" implies a tree-like structure with branches arranged in a specific spatial layout and a predetermined distribution of branch width and length. This fractal model is intended to serve as a foundation for a physical model and a starting point for further research.

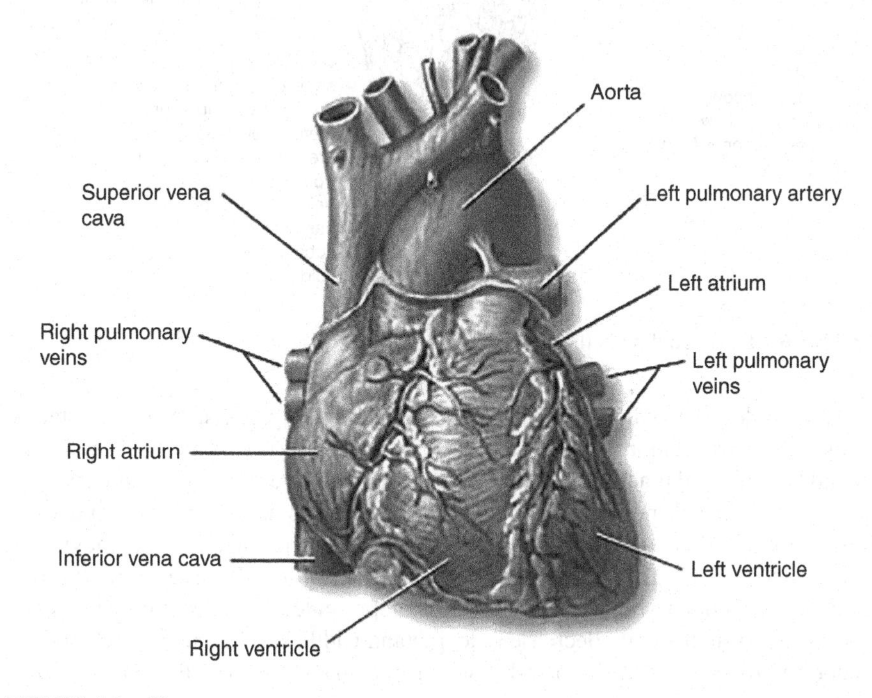

FIGURE 6.9 Heart.

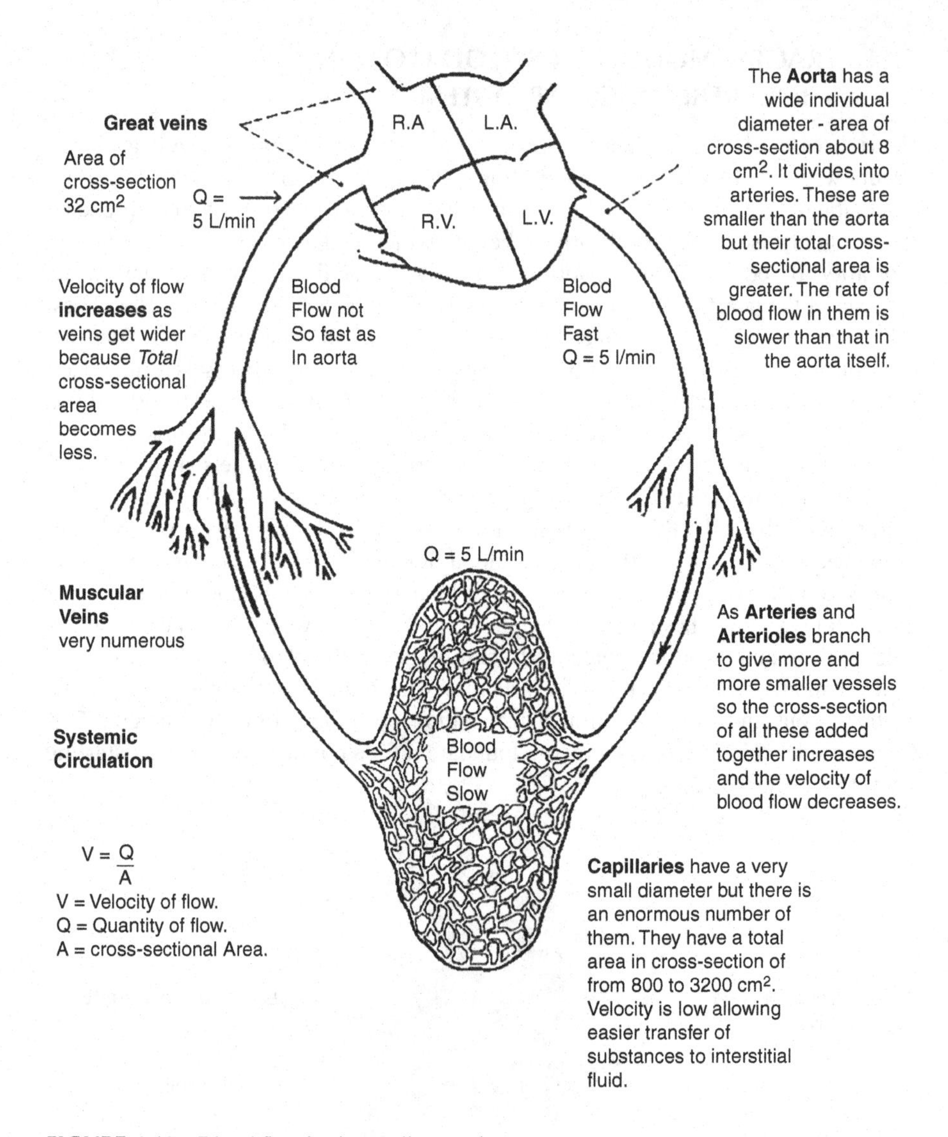

FIGURE 6.10 Blood flow in the cardiovascular system.

The simplest physical principles, such as flow continuity, pressure, and resistance, along with their relationships, should be reflected in the fractal model. The model should be self-similar across multiple scales. Although it represents a geometric simplification, a fractal model of the blood vessel system is adequate for a valid study of blood flow. This analysis facilitates understanding the impact of hemodynamic forces and their role in the emergence of cardiac disorders. From a geometric perspective, the human vascular system can be represented as a complex network of vessels that systematically decrease in length and width while branching irregularly. According to our model, the blood vascular system's Hausdorff–Besicovitch (D_{HB})

and box counting (D_B) dimensions of the blood vascular system match the dimension (D) of the embedding space.

$$D_B = D_{HB} = D = 3 \tag{6.4}$$

Gaining insight into the functioning of the heart can be enhanced by the ability to forecast the flow and pressure at any point in the circulatory system. Flow analysis in the cardiovascular system, using fractal models, can be based on the fractal dimension and fractal geometry of the vascular tree. Elzbieta Gabrys, Marek Rybaczuk, and Alicja Kedzia's work describes blood flow simulation using fractal models of the circulatory system; 3D numerical simulations based on the Navier-Stokes equations model blood flow within this system. However, these outcomes can be costly. This study utilizes permeability and pressure drop in accordance with Darcy's law to propose intricate geometrical (fractal) models of human blood arteries. The Cayley tree has been employed to classify and analyze fractal blood arteries, which are the focus of this work. This article details the fractal analysis of a blood artery located in any region of the cardiovascular system. Moreover, from a computational perspective, numerical simulations of the 3D tree based entirely on Poisuellie's equations are now feasible, thanks to the application of fractals. Figure 6.11 presents a three-dimensional representation of the fractal vascular tree.

Hemodynamics is the area of biophysics that involves the mathematical and experimental analysis of blood's physical characteristics as a fluid in vessels. It encompasses the study of the forces involved in blood flow and how blood moves through the circulatory system. In a closed system of branching vessels, blood circulates. This discussion covers the basic principles of blood flow through arteries, with particular attention to the interactions between pressure, flow, and resistance. Generally, liquids cannot be compressed; thus, a type of flow known as incompressible flow occurs when the fluid's density remains constant. Therefore, blood is also considered an incompressible substance, where p represents the fluid's density and is a constant for incompressible flow. In 1983, Wilkinson and Willemsen developed invasion percolation, a novel version of percolation theory that considers fluid transport mechanisms. Invasion percolation can be used to model blood flow in the capillary bed, where the foundational hypothesis involves mapping the threshold pressure (pc) for each pore (branch of the tree). The Cayley tree is a system in which the percolation problem can be rigorously solved. ; it is also known as the Bethe lattice. The advantage of the

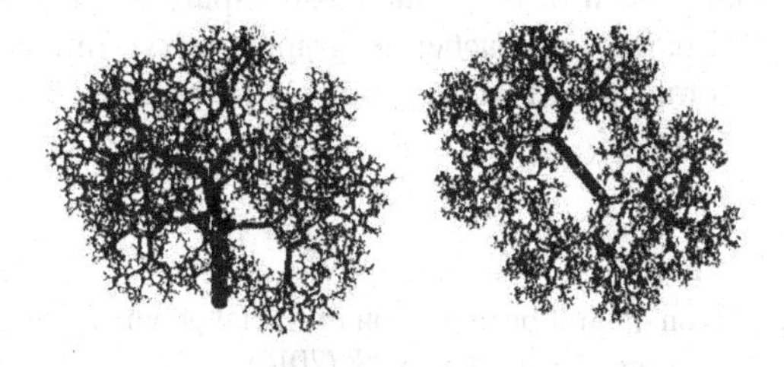

FIGURE 6.11 Three-dimensional geometry of fractal vascular tree.

Cayley tree is that the percolation threshold can be identified, wherein the critical concentration is less than one, allowing for exploration of the regime above it. The blood flow that can percolate from one branch to another is referred to as the percolation threshold. This concept relies on the mapping of each pore's threshold pressure to an occupation probability. We have modeled an approach for traversing the blood vessels and have briefly reviewed numerical computations based on a synthetic view of patients with both aberrant and normal conditions.

6.15 CONCLUSION

With ramifications for numerous scientific and practical fields, the use of fractals in graph theory provides a rich and productive environment for investigation and discovery. By fusing the analytical rigor of graph theory with the mathematical elegance of fractals, scientists can reveal hidden patterns, clarify underlying concepts, and create cutting-edge solutions to address the challenges of an interconnected world. In summary, the utilization of fractals in graph theory presents an intriguing convergence between two disparate disciplines, offering fresh perspectives and insights into the composition and dynamics of complex networks. Fractals provide a potent framework for analyzing and modeling a wide range of phenomena, from biological networks and urban infrastructure to social networks and communication systems, all viewed through the lens of graph theory. The self-similar and hierarchical features of networks can be characterized using fractal-based representations, enabling researchers to identify structural elements crucial to network resilience and dynamics, such as hubs, clusters, and scale-free behavior. Techniques for encoding and compressing data based on fractals offer efficient ways to represent and transfer data within networks. By leveraging the self-similar structure of fractals, researchers can develop new coding methods and protocols that enhance data compression, error correction, and channel capacity in communication systems. Fractal-inspired algorithms and models provide innovative approaches for constructing and refining network structures. By harnessing the efficiency and self-similarity characteristics of fractals, researchers can create scalable and reliable network topologies for various applications, including computer networks, telecommunications, and transportation networks.

Fractal-based metrics and algorithms allows researchers to gain deeper insights into the resilience, organization, and connectivity of spatially distributed systems. This knowledge can be applied to environmental management, infrastructure development, and urban planning. The intersection of fractals and graph theory fosters interdisciplinary collaboration in fields such as computer science, biology, physics, and sociology. Fractal-based approaches in graph theory stimulate creativity and ingenuity in solving complex real-world problems by bridging theoretical concepts with practical applications.

REFERENCES

1. Garg, A., A review on natural phenomenon of fractal geometry, *International Journal of Computer Applications*, Vol. 975, p. 8887 (2014).
2. Ankenbrand, T., and Tomassini, M., Multivariate time series modelling of financial markets with artificial neural networks, In: *Artificial Neural Nets and Genetic Algorithms*, Springer, Vienna, pp. 257–260 (1995).

3. Telcs, A., Spectra of graphs and fractal dimensions, *Journal of theoretical Probability*, Vol. 8, pp. 77–96 (1995).
4. Andres, J., and Rypka, M., Fuzzy fractals and hyperfractals, *Fuzzy Sets and Systems*, Vol. 300, pp. 40–56 (2016).
5. Mandelbrot, B., *The Fractal Geometry of Nature*, Vol. 1, WH Freeman, New York (1982).
6. Bohdalova, M., and Gregus, M., *Markets, Information and Their Fractal Analysis, E-Leader*, CASA, New York, pp. 1–8 (2010).
7. Blackledge, J. M., Evans, A. K., and Turner, M. J., *Fractal Geometry: Mathematical Methods, Algorithms, Applications*, Elsevier (2002).
8. Berry, M. V., Lewis, Z. V., and Nye, J. F., On the Weierstrass Mandelbrot fractal function, *Proceedings of the Royal Society of London. A. Mathematical and Physical Sciences*, Vol. 370(1743), pp. 459–484 (1980).

7 Matching and Its Applications in Real Life

7.1 INTRODUCTION

Regarding graph theory, a matching is a set of non-adjacent edges without any shared vertices. Each of these edges connects to a unique vertex. In other words, a network where each node has either one or zero edges connected to it is referred to as a matching. It is important to distinguish between graph matching and graph isomorphism. While graph isomorphism determines if two graphs are identical, matchings are specific subgraphs within graphs. Graph matching confirms the internal structure of the graph. A matching M of graph G is considered maximal if it is impossible to add an edge that is not already included in M [1]. Including such an edge would violate its status as a matching. In simpler terms, there is no matching of G that can be considered a subset of a maximal matching M. It's important to note that a maximal matching might not always correspond to the subgraph with the greatest number of matches.

Every perfect matching graph is also a maximum matching graph, since no edges can be added to a perfect matching without violating its definition. However, the maximum matching of a graph does not need to be perfect. If |V (G)| has a perfect match, then graph "G" must have an even number of vertices. If the degree of a vertex is odd, one vertex will remain unpaired after the remaining vertices couple with each other, which contradicts the concept of perfect matching [2].

Figures 7.1–7.3 provide examples of matching.

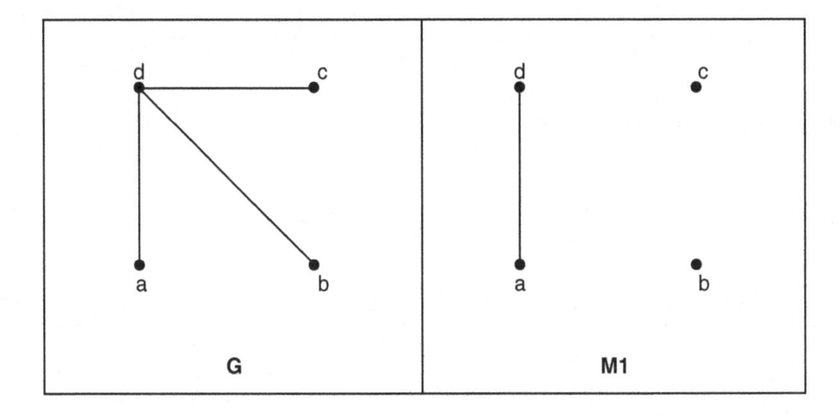

FIGURE 7.1 Matching

DOI: 10.1201/9781003481096-7

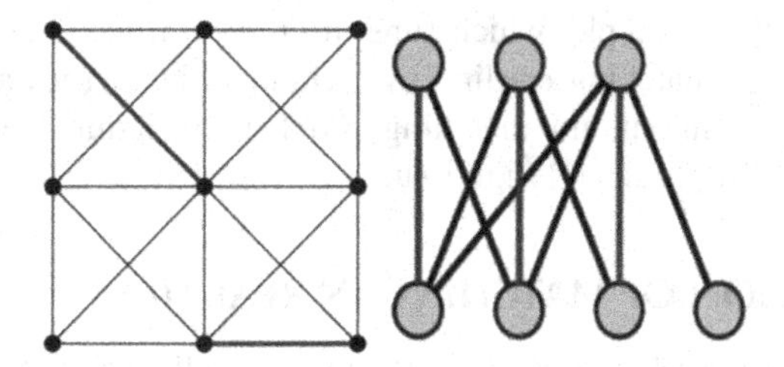

FIGURE 7.2 Maximum matching edges in Hamiltonian graph and bipartite graph

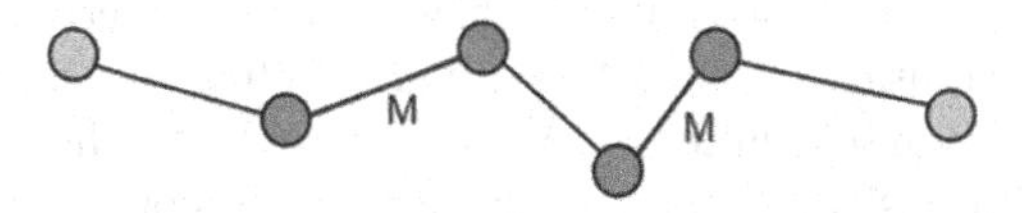

FIGURE 7.3 An alternating path between two vertices like saturated and unsaturated vertices

7.2 RESEARCH AREA OF MATCHING IN GRAPH THEORY

The goal of graph matching is to minimize or maximize the disputes (affinities) between the nodes and edges by identifying the best correspondences between them. This fundamental problem in computer science is mainly related to fields such as multimedia, computer vision, combinatorics, and pattern recognition. Although graph isomorphism is frequently studied in theoretical contexts, inexact weighted graph matching receives greater emphasis due to its flexibility and practical applications.

Numerous reviews of recent research on (inexact) weighted graph matching have appeared in research papers detailing the algorithms, formulations, and methodologies employed. These publications categorize methodologies under several major headings, including the use of different types of graphs, modeling affinities, exploring problem orders, and the matching procedures, among other topics.

For many reasons, the perfect matching problem is crucial to complexity theory, algorithms, and combinatorics. The issue and complexity theory are closely related. The first polynomial time approach for locating the maximum matching in a graph was provided by Edmonds. Interestingly, Edmonds was inspired by this algorithm to suggest polynomial time as a gauge of effective computing. The issue becomes even more intriguing when considering parallel algorithms, as determining which tasks are efficiently parallelizable is a key question in the theory of parallel processing. Most outcomes in bipartite matching can be demonstrated using flow theory within the context of network flows. It was discovered that non-bipartite graphs present a significantly more difficult matching problem. Over a decade later, Edmonds found the first efficient method to identify a maximum cardinality matching in a non-bipartite graph.

O (n 5/2) time is needed to solve the algorithm's broad graph problem. An approximate maximum cardinality matching technique that surpasses O (e+n) sequential time is proposed in the research paper, resulting in a matching of at least e/(n−1). For every bipartite graph, this technique produces a matching size of at least 2e/n. The derived lower bounds are existentially tight, and these procedures are incredibly simple. The weighted example, which runs in O (e+n) time, is extended using the suggested maximum matching cardinality technique. Numerous applications exist for the approximate maximum matching problem, including vertex cover, TSP, MAXCUT, and VLSI physical design issues.

7.3 APPLICATION OF MATCHING IN REAL LIFE

There are numerous applications for matching, including neural networks, scheduling and planning (such as job seeking and arranging machines and jobs), graph coloring (including tasks like finding sick cells, diseases related to blood groups, and jewelry designing), marriage assembly problems, and flow networks (such as tower networks, cell phone usage, Google Maps, and traveling salesman problems). One specific application is in scheduling, where m jobs can be finished by n workers. In a bipartite network, the tasks and workers represent two sets of vertices; a job is associated with a worker if and only if the worker can complete it. Thus, the issue at hand is determining a maximum matching [3].

7.3.1 Matching in Blood Samples

If a person is severely injured and requires a blood transfusion or transplant, the doctor analyzes the patient's blood type and performs cross-matching. If the recipient's blood is compatible with the donor's blood, the next steps can proceed. Blood typing reveals the type of blood the patient has. Analyzing blood type involves identifying specific antigens on the recipient's red blood cells (RBCs). Proteins known as antigens can trigger the recipient's immune system to create antibodies against incompatible blood types. Humans typically have four blood types: Type A, which contains A antigens; Type B, which contains B antigens; Type AB, which contains both A and B antigens; and Type O, which lacks both A and B antigens.

7.3.2 Translation of Blood Matching

The doctor determines which type of blood is suitable for the patient by analyzing the blood type. Some blood antibodies trigger immune reactions against other blood types. AB type blood is referred to as the universal recipient, as this type can receive blood from all other blood groups. O type blood, known as the universal donor, can be provided to individuals of any blood group.

7.4 STAR MATCHING

The Janma Rashi and birth stars of the couple getting married are used to match them for marriage. Before tying the knot and spending the rest of their lives together, the boy and girl in an Indian Tamil marriage match make a crucial decision. Tamil and Hindu siddhas ensure that the couple's compatibility with these important lifelong bonds allows them to live the best possible lives together. Star matching, essentially birth star matching, reveals the quality of marital interaction, known as Dina Porutham and Nakshatra Porutham. Horoscope matching is referred to as Jathaga Porutham. It is one of the fundamental interactions for marriage. It analyzes 13 types of poruthangal or characteristics between two horoscopes and provides marriage reports. This process is used for matching the horoscopes of both the male and female for their relationship and romance. For example, Revathi Nakshatra is generally instinctively compatible with Barani Nakshatra. Hastam, Swati, Shravana, Karthigai, and Pooram Na also show second-level compatibility with Barani Nakshatra. Conversely, it is incompatible with Chitra, Vishaka, Dhanistha, and Purvabhadrapada when paired with Barani Nakshatra.

7.5 MATCHING IN FRACTAL GRAPHS

A collection of graph's non-adjacent edges is called a matching. If a matching set has the greatest number of non-adjacent edges, it is referred to as the maximum matching set. The superset of matching sets that forbids having more than one edge in the matching set is known as the maximal matching set. In terms of the matching set, it has the fewest edges while enclosing the adjacent edges throughout the entire graph. The number of edges in the matching set is known as matching cardinality, denoted as M(G). This typically contains around half of the graph's edges. If each graph point is M-saturated, the matching set is referred to as a perfect matching. Let M be a matching set of graph G. If a path in G has edges that alternately fall in X-M and M, it is referred to as an M-alternate path. If both the origin and terminal vertices of an augmenting path are unsaturated, the path is referred to as an alternative path. The maximum matching cardinality is the total number of edges in a matching set [4].

7.6 TYPES OF MATCHING

Matching is the most developed area of graph theory. It involves the collection of non-touching nodes in a graph. Vertices in this collection are definitely non-adjacent to one another, while simultaneously being adjacent to other nodes in the graph. The collection of vertices in the matching set and the non-matching set are represented as a function, which is referred to as a bipartite graph. There are three types of matching in matching theory: maximum or largest matching, maximal or minimal matching, and perfect matching. Largest matching consists of non-adjacent edges, meaning that the collection of edges does not share any common vertices among themselves. Largest matching cardinality M(G) is defined as the collection of these non-adjacent edges. These edges must be adjacent to the non-matching vertices in the given graph. Matching M′ is said to be maximal if adding any edges is not possible in M′. If it is done, the set makes M′ as non-matching set. This kind of set is considered by M′ (G). A matching is called perfect matching if the collection of edges is adjacent to all other vertices in the graph; therefore, all vertices are matched and belong to the matching set.

7.7 MATCHING IN VARIOUS GRAPHS

- **Null Graph:** A graph with a single vertex and no edges is called a null graph. The matching set is the selection of non-adjacent edges in the graph. It is concluded that both the trivial graph and the null graph have no matching (either maximum or maximal) cardinality.

- **Trivial Graph:** A single vertex graph is referred to as a trivial graph. Since a trivial graph has no edges, its matching set also contains no edges. Therefore, both maximum or largest matching cardinality and maximal or minimal matching cardinality are zero.
- **Regular Graph:** A graph is referred to as regular if every vertex has the same degree. The degree of a regular graph is determined by its cardinality. A graph is m-regular if each vertex has a degree of m. For all regular graphs, the above formulas 7.1 and 7.2 are used to calculate the maximum matching cardinality and the maximal matching cardinality. The formulas are as follows:

$$\text{Maximum or largest matching cardinality} = \frac{V}{2} \tag{7.1}$$

$$\text{Maximal matching cardinality} = \frac{V}{2} \tag{7.2}$$

where V is vertices of the graph.

(Tables 7.1–7.3).

$$\text{Maximum matching cardinality} = \frac{2}{2} = 1,$$

$$\text{Maximal matching cardinality} = \frac{2}{2} = 1.$$

TABLE 7.1
Maximum Matching Cardinality and Maximal Matching Cardinality for Two-Regular Graph

S. No.	Degree of the Graph (m)	Number of Vertices V(G)	Two-Regular Graph	Maximum Matching Cardinality	Maximal Matching Cardinality
1	2	3	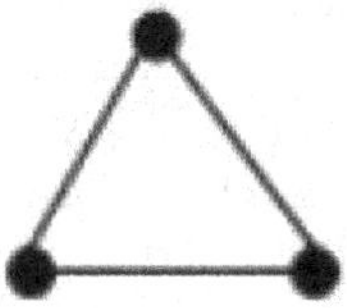	1	1
2	2	4	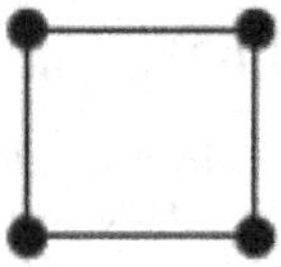	2	2

(Continued)

TABLE 7.1 (*Continued*)
Maximum Matching Cardinality and Maximal Matching Cardinality for Two-Regular Graph

S. No.	Degree of the Graph (m)	Number of Vertices V(G)	Two-Regular Graph	Maximum Matching Cardinality	Maximal Matching Cardinality
3	2	5	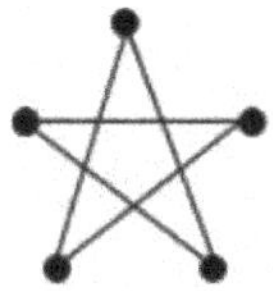	2	2
4	2	7	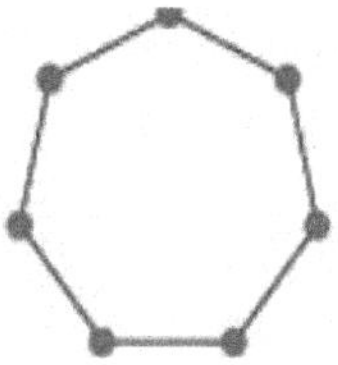	3	3

TABLE 7.2
Maximum Matching Cardinality and Maximal Matching Cardinality for Three-Regular Graph

S. No.	Degree of the Graph	Number of Vertices V(G)	Three-Regular Graph	Maximum Matching Cardinality	Maximal Matching Cardinality
1	3	4	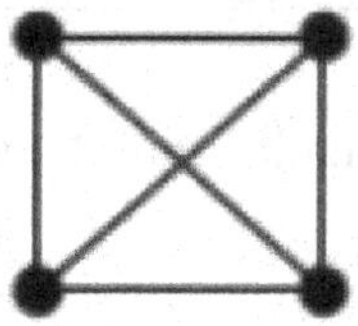	2	2
2	3	4	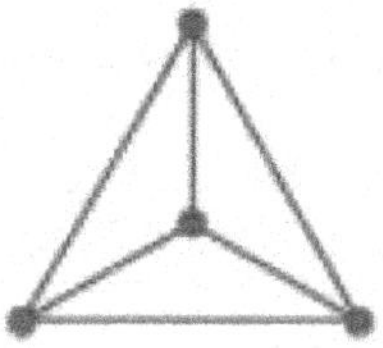	2	2
3	3	5	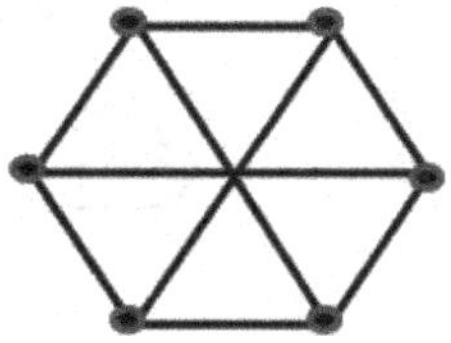	3	3

(*Continued*)

TABLE 7.2 (*Continued*)
Maximum Matching Cardinality and Maximal Matching Cardinality for Three-Regular Graph

S. No.	Degree of the Graph	Number of Vertices V(G)	Three-Regular Graph	Maximum Matching Cardinality	Maximal Matching Cardinality
4	3	8	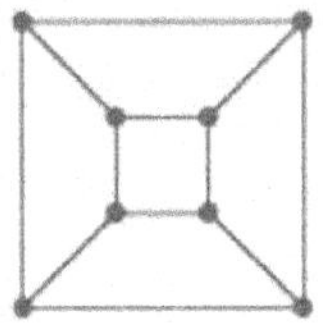	4	4
5	3	8	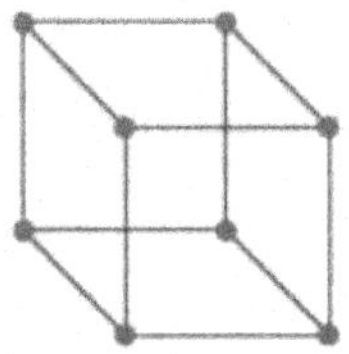	4	4
6	3	8	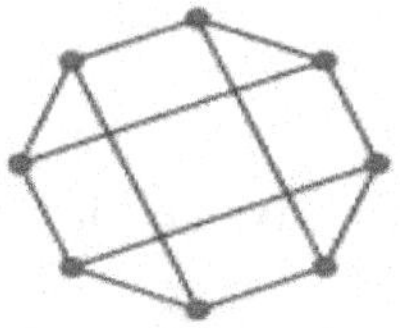	4	4

TABLE 7.3
Maximum Matching Cardinality and Maximal Matching Cardinality for Four-Regular Graph

S. No.	Degree of the Graph	Number of Vertices V(G)	Four-Regular Graph	Maximum Matching Cardinality	Maximal Matching Cardinality
1	4	5	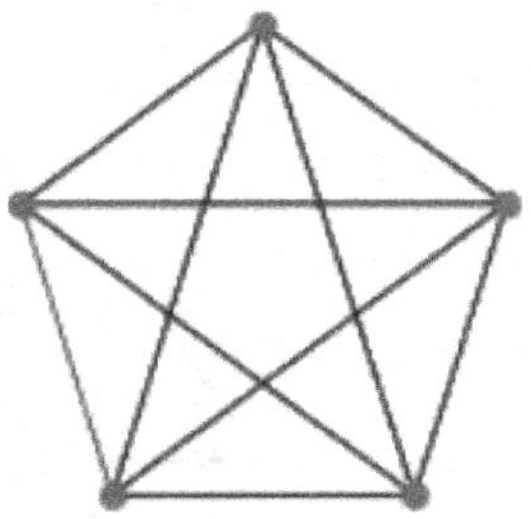	2	2

(*Continued*)

TABLE 7.3 (*Continued*)
Maximum Matching Cardinality and Maximal Matching Cardinality for Four-Regular Graph

S. No.	Degree of the Graph	Number of Vertices V(G)	Four-Regular Graph	Maximum Matching Cardinality	Maximal Matching Cardinality
2	4	8	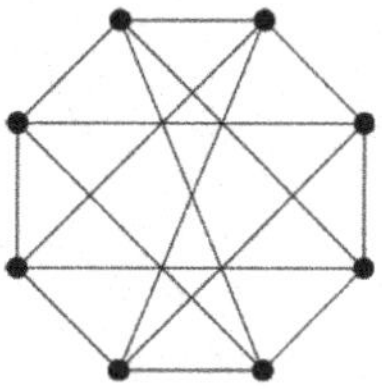	4	4
3	4	7	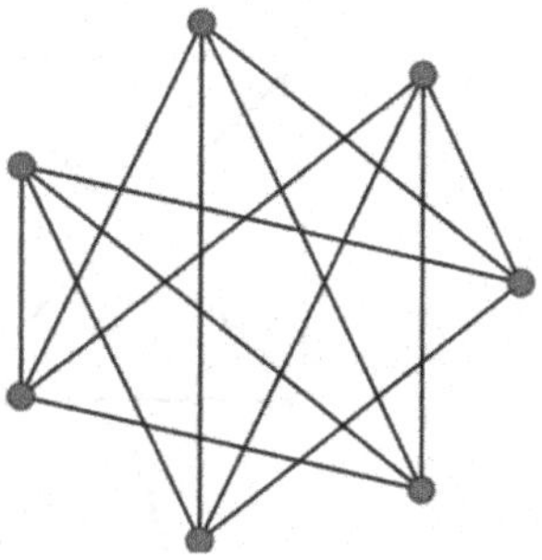	3	3

- **Complete Graph:** Graph: Each vertex in this type of graph is adjacent to all other vertices in the graph. It is denoted as Km, where m represents the number of vertices. Every vertex has a degree of m-1. The graph is both strongly linked and closed. Formulas 7.3 and 7.4 are used to evaluate maximum matching and maximal matching cardinality for all complete graphs (Table 7.4).

The formulas are as follows:

$$\text{Maximum matching cardinality} = \frac{\text{No. of vertices}}{2} \tag{7.3}$$

$$\text{Maximal matching cardinality} = \frac{\text{No. of vertices}}{2} \tag{7.4}$$

- **Cycle Graph:** This graph is a closed, path-connected graph. Its edges and vertices are the same elements. Each node is connected to the others individually via edges in a path-like manner. There is only one path starting

TABLE 7.4
Maximum Matching Cardinality and Maximal Matching Cardinality for Complete Graph

S. No.	Number of Vertices (k)	k-Complete Graph	M(G)	M′(G)
1	2		1	1
2	3	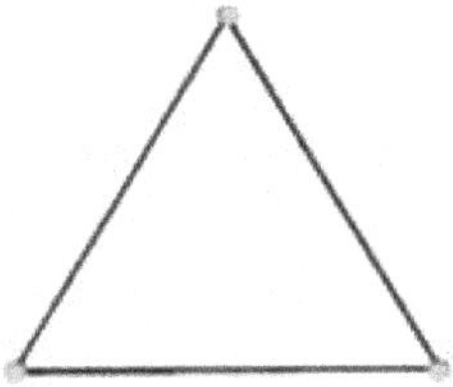	1	1
3	4		2	2
4	5	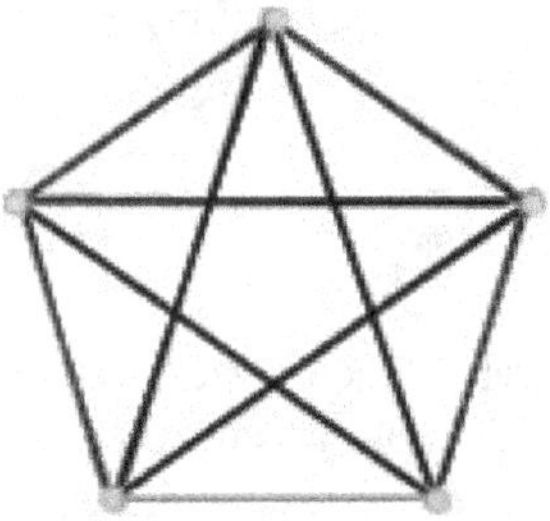	2	2
5	6	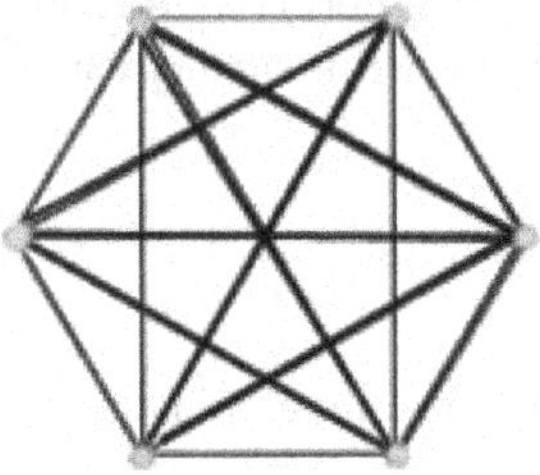	3	3
6	7	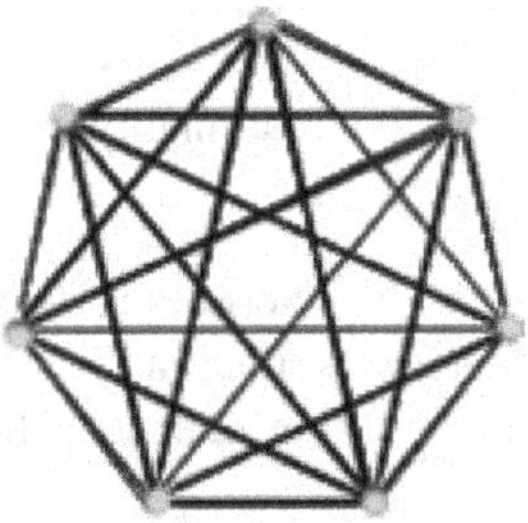	3	3

at one vertex and returning to it. There are no edges that cross over one another. The degree of each vertex is the same, with every vertex sharing a degree of two. In an alternate iteration, an additional edge is added. The cycle graph begins with three edges. For a complete cycle graph, the total number of edges is equal to the number of vertices. It is denoted as Cm, where "m" is the number of vertices in the cycle graph. Maximal and maximum cardinalities are calculated using formulas 7.5 and 7.6 (Table 7.5).

$$\text{Maximum matching cardinality } M(G) = \frac{\text{No. of vertices}}{2} \quad \text{or} \quad \frac{\text{No. of edges}}{2} \tag{7.5}$$

$$\text{Maximal matching cardinality } M'(G) = \frac{\text{No. of vertices}}{3} + 1 \quad \text{or} \quad \frac{\text{No. of edges}}{3} + 1 \tag{7.6}$$

TABLE 7.5
Maximum Matching Cardinality and Maximal Matching Cardinality for Cycle Graph

S. No.	Number of Vertices (n)	Cycle Graph C_n	$M(G)=\frac{n}{2}$	$M'(G)=\frac{n}{3}+1$
1	3		1	1
2	4		2	2
3	5		2	2

- **Star Graph:** The star graph is a connected graph where all edges are adjacent to a single vertex. It can be represented as a function graph, showing a complete Bipartite Graph. In this type of graph, any one of the edges can be selected for the maximum matching set and the maximal matching set. The maximum matching and maximal matching cardinalities are both equal to one.
- **Wheel Graph:** This type of graph is a Closed Loop Graph, resembling the shape of a wheel, hence its name. It is denoted by Wm, where m represents the number of vertices. It has m vertices and 2(m – 1) edges. It has partitioned into boundary vertices and center vertex. All the boundary m – 1 vertices are connected one by one through edges and finally all vertices are connected to center vertex by m – 1 edges. Maximum or largest matching cardinality and maximal or Minimal matching cardinality are calculated in wheel graphs by the following formulas 7.7 and 7.8 (Table 7.6).

$$\text{Maximum matching cardinality} = \frac{\text{No. of vertices}}{2} \tag{7.7}$$

$$\text{Maximal matching cardinality} = \frac{\text{No. of vertices}}{3} + 1 \tag{7.8}$$

TABLE 7.6
Maximum Matching Cardinality and Maximal Matching Cardinality for Wheel Graph

S. No.	Number of Vertices	Wheel Graph	Maximum Matching Cardinality	Maximal Matching Cardinality
1	4	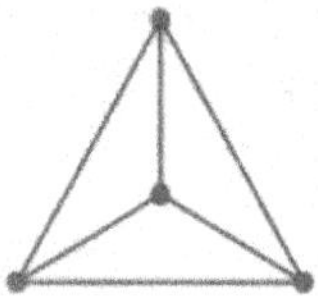	2	2
2	5	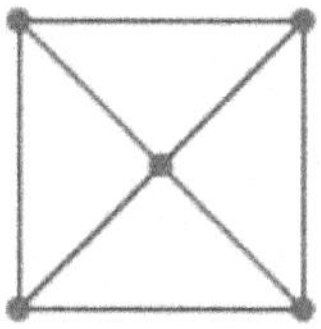	2	2
3	6	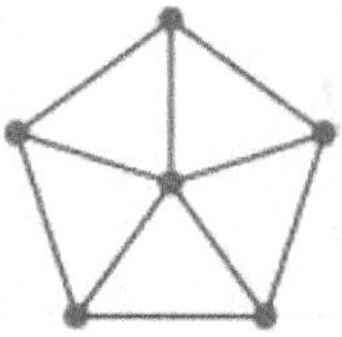	3	2

(Continued)

TABLE 7.6 (*Continued*)
Maximum Matching Cardinality and Maximal Matching Cardinality for Wheel Graph

S. No.	Number of Vertices	Wheel Graph	Maximum Matching Cardinality	Maximal Matching Cardinality
4	7		3	3
5	8		4	3

7.8 APPLICATION OF MATCHING

Matching is an advanced tool in graph theory that has significant implications in everyday life. The world-famous Konigsberg Bridge Problem was solved by Euler, and matching also provides solutions to this problem.

7.8.1 Matching in Konigsberg Bridge Problem

The current condition of the world-famous Konigsberg Bridge is deteriorating. Two bridges are no longer in use due to damage sustained during the Second World War. The other two bridges have been modified in contemporary ways. The remaining three bridges have remained unchanged since Euler's time. Matching is employed to determine which bridge is the most important and useful.

- **Bridges of Konigsberg Facts:** This problem is closely related to the study of mathematics and was solved by Leonhard Euler in 1735. It contributes to the development of graph theory and topology. The city of Konigsberg is situated on both sides of the Pregel River, which was once part of Prussia but is now located in Kaliningrad, Russia. The islands are not connected by any routes other than the bridges. A walk cannot start and finish at the same point. Euler resolved this problem.
- **History of Konigsberg: Konigsberg:** Under the rule of King Ottokar II, the city of Konigsberg was founded by the Teutonic Knights in 1254 after their conflict with the Prussian people. The citizens built seven bridges across the river that connected the two islands, Kneiphof and Lomse. Later, it became a famous trading hub. The area developed into a prosperous locale, where ships and trade contributed to the affluent and comfortable lifestyle of the people of Konigsberg, significantly benefiting local merchants and their families.

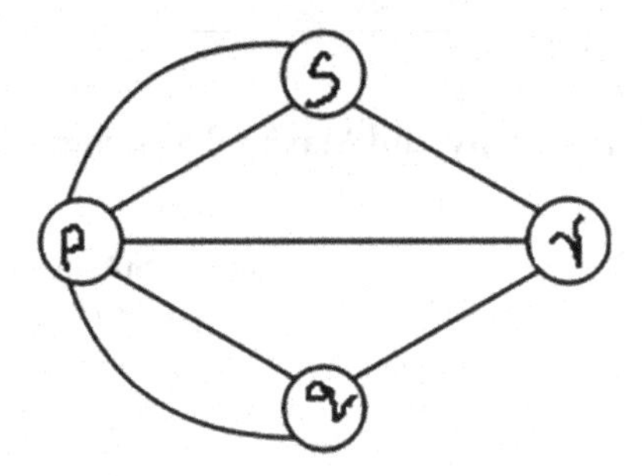

FIGURE 7.4 A visual depiction of the Konigsberg Bridge Issue

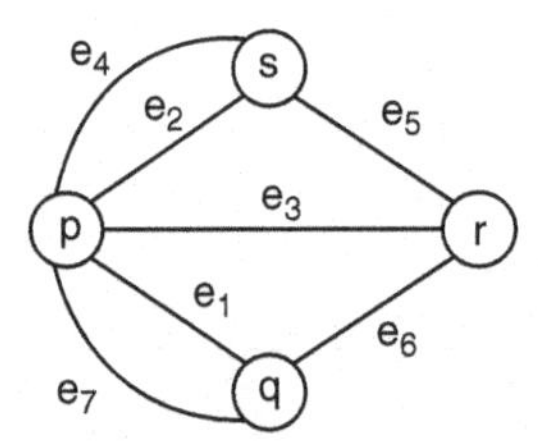

FIGURE 7.5 Formal Representation of Konigsberg Bridge Problem

The world-famous Seven Bridges of Konigsberg are now known as Kaliningrad, located in Russia. It was formerly part of Prussia and sits on both sides of the Pregel River, which separates two distinct, unconnected islands—Kneiphof and Lomse. These islands are connected to each other by the seven bridges. In the 18th century, the famous problem arose concerning the bridges and their geographical layout. The question was whether it was possible for a citizen to walk through all the bridges exactly once. During this time, the problem prompted significant interest in the bridges and their landscape. The challenge was for a person to traverse all the bridges precisely once. No one could provide a solution to this problem until Euler resolved it simply.

Figures 7.4 and 7.5 show the structure and graphical format of Konigsberg Bridge Problem.

7.8.2 Rising of Konigsberg Bridge Problem

Kaliningrad is a seaport established by Soviet revolutionaries. It is located near the Baltic Sea, between Poland and Lithuania. The German Kingdom of Prussia named this city Konigsberg. Nowadays, it has become related to a mathematical puzzle that led to the development of many areas in mathematics, such as topology and graph theory. In the 18th century, the people of Kaliningrad found it difficult to navigate the complex arrangement of bridges. They hoped to minimize their time and expenses.

- **Euler's Analysis:** The edges are connected by certain rules, transforming the depiction of the bridges into a graph. This representation is very useful for understanding the layout of roads and meeting places. Leonhard Euler observed that whenever a person crosses a bridge to reach one vertex and exits another vertex via a bridge, the degrees of the nodes determine whether it is possible to traverse the graph and cross each edge once. The number of edges that connect to a node determines its degree. According to Euler, the network must be connected and contain exactly zero or two nodes of odd degree for such a walk to be possible. Carl Hierholzer later provided proof for the result that Euler had claimed. We now refer to such a route as an Eulerian path or Euler walk.

7.8.3 Solution to Konigsberg Bridge Problem

Konigsberg bridge problem is a historically significant issue in the field of mathematics, particularly in the study of topology. In 1735, Leonhard Euler provided an answer to this problem, concluding that it is not possible to traverse all the bridges exactly once. His findings led to the establishment of a theorem using graph theory. In this context, the four land areas were considered as vertices, and the bridges connecting them were treated as edges. Figure 7.4 presents a visual depiction of the Konigsberg Bridge Problem. Euler proved this problem by examining the degree of each vertex. He established a new theorem based on his solution, which states:

- A node has an even degree iff an Euler circuit is a connected graph
- Any two vertices have an odd degree if and only if an Euler path exists. Based on this concept, it was determined that the Konigsberg Bridge does not have an Eulerian path. If there were no vertices with an odd degree, the graph of the Konigsberg Bridge would be Eulerian.

7.8.4 Matching in Konigsberg

Matching is a crucial concept in graph theory, defined as the selection of non-adjacent edges from a given graph. These edges consist of distinct vertices, meaning no vertex intersects with another. There are three main types of matching: largest matching, minimal matching, and perfect matching. Figure 7.5 presents a formal representation of the Konigsberg Bridge Problem.

7.8.4.1 Calculation of Maximum Matching in Konigsberg Bridge Problem

The land areas of Konigsberg can be represented by vertices, while the bridges connecting these areas are represented by edges. It consists four vertices p, q, r, s and seven bridges E_1, E_2, E_3, E_4, E_5, E_6 and E_7. Maximum matching set = {E_2,E_6} or {E_4,E_6} or {E_1,E_5} or {E_7, E_5}.}. Maximal Matching Set= E_3. p and r vertices have maximum adjacent. p has degree 5. It is definitely including Maximal Matching set. The Bridge of e_3 definitely uses many times. This section outlines potential modifications to the graph to make all bridges traversable in a single journey. The degrees of the bridges in the graph may need to be adjusted

TABLE 7.7
Maximal Adjacent Vertex of Konigsberg Graph

S. No.	Vertices	Degree	Adjacent Vertices	Maximal Adjacent
1	p	5	q, r, s	p
2	q	3	q, r	—
3	s	3	p, r	—
4	r	3	p, r, s	r

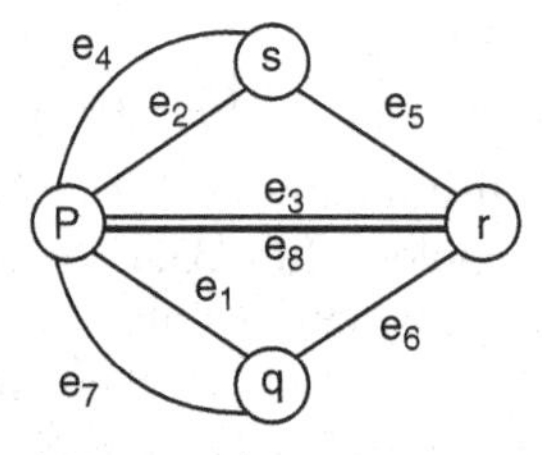

FIGURE 7.6 New Solution of Konigsberg Bridge Problem

TABLE 7.8
Degree of Vertex in Konigsberg Bridge

S. No.	Vertices	Degree	Adjacent Vertices
1	p	6	q, r, s
2	q	3	p, r
3	s	3	p, r
4	r	4	p, r, s

7.8.5 Alternative Solution to the Konigsberg Bridge Problem

7.8.5.1 Case 1

One potential solution to the Konigsberg Bridge Problem is to add a new edge or construct a bridge from A to D. This modification would render the problem solvable and allow the residents of Konigsberg to traverse all the bridges exactly once. The following figure illustrates that the Konigsberg Bridge is ready for traversal. The dark edge in this graph represents the new edge that was added from A to D. Figure 7.6 depicts this new solution to the Konigsberg Bridge Problem (Table 7.8).

The Euler path is {e_5, e_3, e_8, e_6, e_7, e_4, e_2, and e_1}. }. This path is used to traverse all the bridges exactly once. The bridge e8 was additionally built from p to r, changing the degree of each vertex. Vertices p and r are adjacent to each other by parallel edges. The vertex p has the maximum degree. The actual graph of Konigsberg has a maximum degree of five, but the addition of this new edge increases the degree of vertex p by one. The edge e8, which connects p and r, is considered as a maximal matching. This edge serves as a minimal edge covering, adjacent to all the vertices of the given graph.

7.8.5.2 Case 2

Another way to make the graph traversable is by adding one edge or building one bridge from s to q. Now, the graph is ready for traversal.

The Euler path is e_6 e_1-e_2-e_8-e_7-e_4-e_2-e_3.

Maximum Matching set = {e_5, e_8} or {e_6, e_4} or {e_2, e_6} or {e_3, e_8}.

Maximum Matching Cardinality = 2.

Maximal Matching Set = {e_5, e_8} or {e_6, e_4} or {e_2, e_6} or {e_3, e_8}.

Maximal Matching Cardinality = 2

This additional edge equalizes the cardinality of the maximum matching and maximal matching sets. All vertices are included in the matching set, indicating that this graph has a perfect matching. There are two bridges that are used multiple times in this graph (Table 7.9).

The additional edge is represented by the dark edge in Figure 7.7.

- **Conclusion of Konigsberg Bridge Problem:** This additional edge equalizes the cardinality of the maximum matching and maximal matching sets. All vertices are included in the matching set, indicating that this graph has a perfect matching. There are two bridges that are used multiple times in this graph.

TABLE 7.9
Adjacent Vertices of New Solution of Konigsberg Bridge

S. No.	Vertices	Degree	Adjacent Vertices
1	p	5	q, r, s
2	q	4	p, r, s
3	s	4	p, q, r
4	r	3	p, q, r

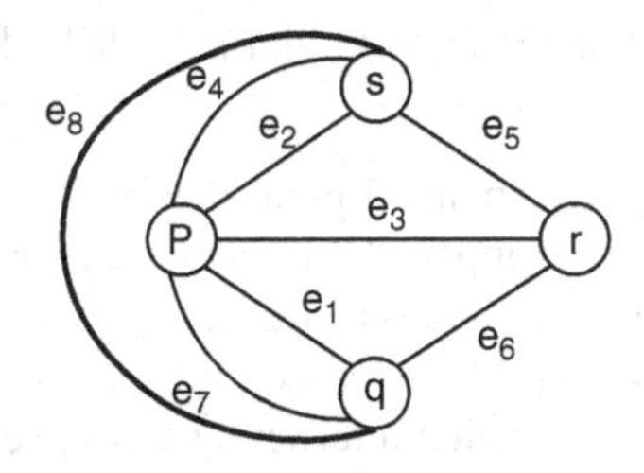

FIGURE.7.7 Alternative Solution of Konigsberg Bridge Problem

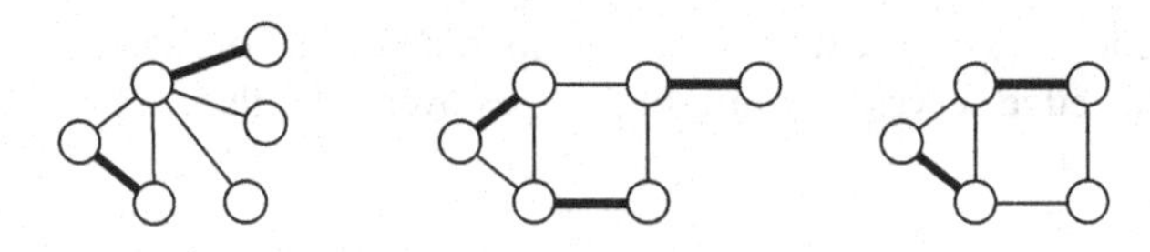

FIGURE 7.8 Matching Set

7.9 MAXIMUM MATCHING IN FRACTAL GRAPHS

The graph theory includes a modern study known as matching, which has several useful applications in real-life problems. Nowadays, many mathematicians are focusing on the research area of matching. In matching, non-adjacent edges do not share a common vertex. The nodes of the matching set are referred to as saturated or impregnated, while the other vertices are called unsaturated. Each selected edge is incident with other edges in the given graph.

Examples are shown in Figure 7.8.

The vertices of the red edges are saturated. These selected non-adjacent edges have no common vertices between them. In this chapter, the maximum matching cardinality is calculated in various fractal graphs such as the von Koch curve, Koch snowflake, Sierpinski arrowhead curve, Sierpinski triangle, and Hilbert curve. Iterative methods and the mathematical induction method are used to calculate the maximum matching cardinality in this chapter. Theorems are derived using the induction method for all natural numbers. Maximum matching is defined as the compilation of the maximum number of non-adjacent lines selected, denoted by M (G).

Theorem 7.1

Matching Set is a called as Maximum Matching Set if G's path is not expanded

PROOF:

Let K be a maximum matching in G. Take the path $P=(v_0, v_1, v_2, \dots, v_{m+1})$. Expanded path the lines $v_0v_1, v_2v_3, \dots, v_{2k}v_{2k+1}$ are not in K, and the lines v_1v_2, $v_3v_4, \dots, v_{2m-1}v_m$ are in K.

Hence, $K' = \{v_1v_2, v_3v_4, \dots, v_{2m-1}v_m\}$ U$\{v_0v_1, v_2v_3, \dots, v_{2k}v_{2k+1}\}$ is a collection of edges in the matching set in G and $|K'| = |K| + 1$. This contradicts the assumption.

Conversely, if G has a non-expanded path, then a matching set M has maximum matching cardinality. Suppose G has a non-expanded path. If M is not a maximum matching in G, there exists a matching M′ of G such that $|M'| > |M|$. Let $I = [M\Delta M']$. Each part of M has two options. Every portion of I is a cycle with an even number of edges that alternately lies in set M and set M′. In the second thought, path P edges alternately lie in M and H′.

7.10 FRACTALS

A fractal is a rough, choppy, or crumbled geometric pattern that is subdivided into smaller parts. Each part resembles a copy of the whole structure. The term "fractal" originates from the Latin word "fractus," meaning "broken." It serves as a general name for a varied class of geometrical objects and typically has the following properties:

- Fractals is the edge structure with different scales
- Fractals may be regular or irregular that supposed with classical Euclidean geometry
- Fractals are formed by the property of self-similarity.
- Hausdorff dimension of fractal structure is strictly better than one dimension.
- Fractals have a straightforward and transparent definition. They can be defined recursively.

This chapter analyzes the maximum matching set and maximum matching cardinality of various fractal graphs, such as the Cantor set, Von Koch curve, Koch snowflake, Sierpinski triangle, Sierpinski arrowhead curve, star fractal graph, and Hilbert curve.

7.11 EVALUATION OF MAXIMUM MATCHING IN FRACTAL GRAPHS

In this section, the maximum matching cardinality of self-similar fractal graphs can be calculated. This section also explains their structure and the implementation of vertices and edges for all iterations.

7.11.1 Cantor Set

The Cantor set is defined by a single line at the beginning stage with two vertices. Georg Cantor used the Cantor set in his paper, implementing it into a famous fractal graph. It is built by a subsidiary rule. The set comprises a single edge segment that passes through it. It was discovered by John Smith but implemented by Georg Cantor. In the first iteration, the interval causes the single line to be divided into three sections: (0, 1/3) U (1/3, 2/3) U (2/3, 1). The middle part is eliminated, partitioning the set into two edges identified by the intervals (0, 1/3) and (2/3, 1). The process of the first iteration is again applied to the remaining two edges. Now, these two edges are split into four edges with the intervals [0, 1/9] ∪ [2/9, 1/3] ∪ [2/3, 7/9] ∪ [8/9, 1], eliminating the intervals (1/9, 2/9) and (7/9, 8/9). This same process continues in the next level of iteration.

Figure 7.10 is the initial iteration of the Cantor set.

Figure 7.11 is the first iteration of the Cantor set.

Figure 7.12 is the second iteration of the Cantor set.

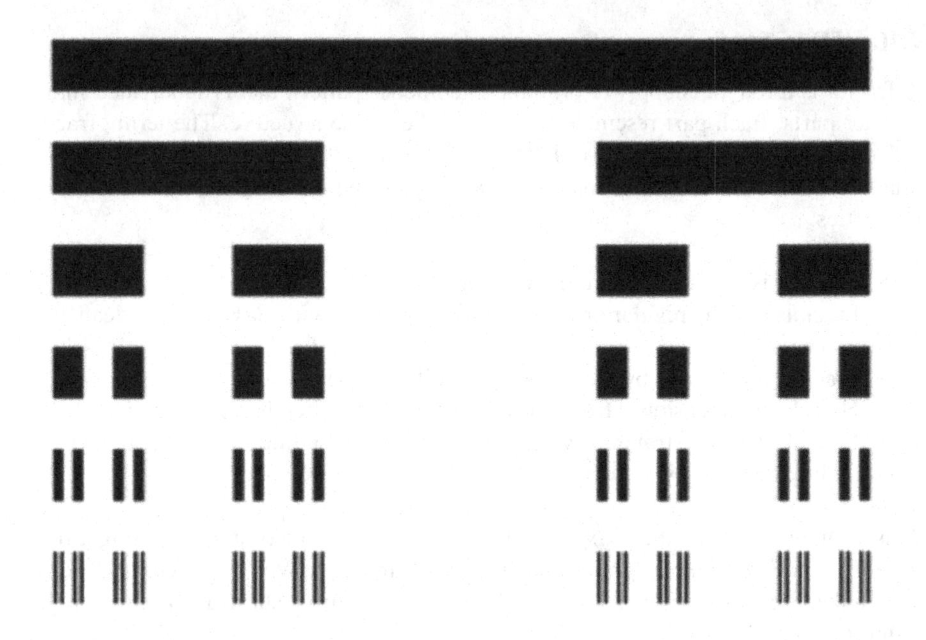

FIGURE 7.9 Cantor set

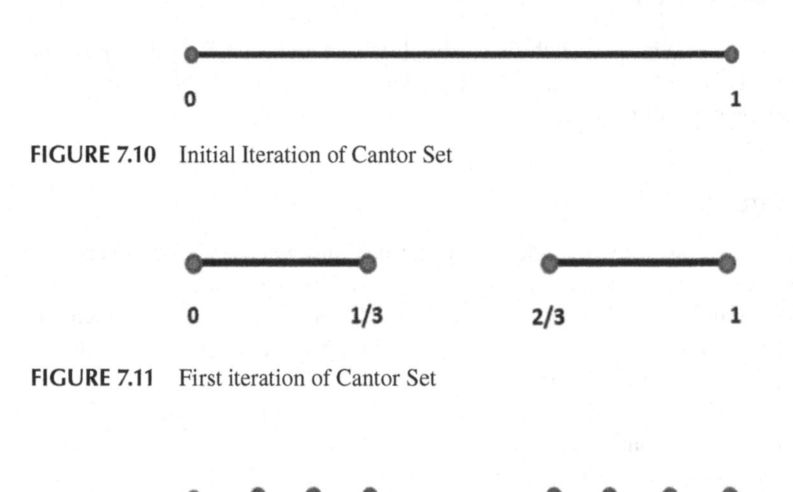

FIGURE 7.10 Initial Iteration of Cantor Set

FIGURE 7.11 First iteration of Cantor Set

FIGURE 7.12 Second iteration of Cantor Set

7.11.1.1 Number of Edges in Cantor Set

Count of vertices and edges in the Cantor set is calculated using the iteration method. In the initial iteration, the Cantor set has a single edge with two vertices in the interval [0, 1].

Iteration 1: In Iteration 1, the unit interval [0, 1] can be partitioned into three parts with scale 1/3. Each part of Cantor Set has length (1/3). This process starts with an elimination method. The interval [0, 1] loses the open middle third (1/3, 2/3). The two line segments that are left are [0, 1/3] and [2/3, 1]. Therefore, this iteration has two non-adjacent edges and four vertices.

Iteration 2: Predecessor iteration are partitioned into three parts. Each part has length 1/9. Removing the open middle part edges (1/9, 2/9) and (7/9,8/9) from the open interval, the resulting line segments are [0, 1/9] ∪ [2/9, 1/3] ∪ [2/3, 7/9] ∪ [8/9, 1]. In total, there are four non-touching edges and eight vertices. The same process continues in the subsequent iterations.

Iteration n: For the nth iteration, the single edge has 2^{n-1} non-adjacent edges and 2^n vertices where n is a whole number and increased by one.

7.11.1.2 Cantor Set

Maximum matching is calculated in the Cantor set. The matching set considers the non-adjacent edges of the graph. In this case, the Cantor set consists solely of non-adjacent edges across all iterations. Therefore, all of the edges of the Cantor set are included in the matching set. The maximum matching cardinality for the Cantor set in the related iterations is equal to the total number of edges enclosed (Table 7.10).

7.11.2 Maximum Matching Cardinality in von Koch Curve

In 1904, the von Koch curve was introduced. The researcher presented a paper on Fractal Geometry, proving that some figures are continuous everywhere but nowhere differentiable. The theme of the paper provides a geometric construction based on the Koch curve, which is analytically expressed in a specific function. This work

TABLE 7.10
Maximum Matching Cardinality in Cantor Set

Iteration n	No. of Edges	Maximum Matching Cardinality	Diagram
1	1	1	1
2	2	2	1/3
3	4	4	1/9
4	8	8	1/27
5	16	16	1/81

contradicted the concepts of Karl Weierstrass. The researcher demonstrated many asymmetrical fractal images found in nature. Later, these concepts were utilized by Mandelbrot in his foundational work on Fractals [4].

7.11.2.1 Construction

This section describes one of the most famous early Fractal Graphs. Its construction is notably interesting. It begins from a topological perspective, meaning one dimension. It starts with a simple edge represented by a unit line. This line is partitioned into three equal parts using a common scale. The middle segment is then partitioned into two lines, which are moved upward and bent. This process is repeated across all the edges in subsequent levels, a process referred to as iteration. The diameter increases significantly in the next iteration. Generally, the edges of the von Koch curve are divided into four parts, with the size of the segments further divided into three. In alternating iterations, the length increases by a factor of 4/3. The Cantor set has 2n edges in the nth iteration. Here, n is even value. It is started at 0 and also incremented by 2. The initial level of the von Koch curve begins with a single edge and two vertices. At the initial stage, this transforms into four edges with five vertices. In the second stage, this expands to 16 edges with 17 vertices. In general, all repetitions of the von Koch curve have an even-length path, except for the initial iteration, which has a different topological dimension. n^{th} iteration of Von Koch Curve has 2^n edges and 2^n +1 vertices itself. Here, n is a whole number that starts at 0 and increases by 2 only.

7.11.2.2 Calculation of Maximum Matching Cardinality Using Mathematical Induction Method

The calculation of the cardinality of the maximum matching set has also been implemented using a theorem. Theorem 7.2 demonstrates the evaluation of the maximum or largest matching cardinality by using the mathematical induction method. This method is a well-known technique in mathematical proofs. It essentially proves that the statement for any theorem holds for all natural numbers n = 0, 1, 2, ... This means that the statement is a sequence of infinitely many cases of n.

Theorem 7.2

Let Von Koch Curve has 2^n edges in the nth iteration. Then it has 2^{n-1} edges in the nth iteration for all n is a natural number.

PROOF:

Theorem used an induction method. Let n=2. In the initial level of Von Koch Curve has 2^2 edges. Alternative edges are considered for Maximum or Largest Matching Set. It has Maximum Matching Cardinality is 2 only. Assume that statement is accept for all n=k. It means that, k^{th} level of Von Koch Curve has 2^{k-1} edges for Maximum Matching Set.

TO PROVE:

k+1th iteration of Von Koch Curve have has 2^k edges in Maximum Matching Set. k+1th iteration of Von Koch Curve has 2^{k+1} edges. It has subdivided into two sets with 2^k number of edges and 2 numbers of edges. k^{th} iteration of Von Koch Curve of set 1 has 2^{k-1} edges in Maximum Matching Set by the assumption. Remaining 2 edges of set 2 are connected to set 1 by using combination method. Therefore, k^{th} iteration of Von Koch Curve has $2^{k-1} * 2 = 2^k$ number of edges. Induction method is verified for all real number. Thus the theorem is proved for all natural number.

7.11.2.3 Calculation of Maximum Matching Cardinality Using Iteration Method

This method involves the selection of non-touching edges for the matching set. Here, the maximum number of non-adjacent edges is considered at higher levels of iteration of the von Koch curve. It may not be clearly visible and can be difficult to discern. Alternate edges are selected for the largest matching set, but this is not an augmented path. An augmented path is a path that does not include terminal edges. Therefore, alternate edges in succession are not considered for the maximum matching set. At the initial stage of the von Koch curve, there is a single edge of odd length, indicating that it has a specific topological dimension. Consequently, the given formula is not satisfied at this early stage. In subsequent iterations of the von Koch curve, the formula is satisfied, even when it has an even path length. The cardinality of the maximum matching is evaluated using an iterative method. Here, the common generating formulas are checked against the alternative stages of the given fractal graph, which is an open-path fractal graph. Generally, all open-path fractal graphs follow a common formula for maximum matching cardinality.

$$M(G) = \frac{\text{Count of edges}}{2} \text{ if count of edges is even} \tag{7.9}$$

In Iteration 1 of the von Koch curve, the maximum matching cardinality is 2. In Iteration 2 of the von Koch curve, the maximum matching cardinality is 8. This review of the previous iterations 1 and 2 demonstrates that for every iteration of the von Koch curve, the maximum matching cardinality equals half of the number of edges. Ultimately, it is determined that the semi-edges for the corresponding iteration represent the maximum matching cardinality. It can be tabulated in Table 5.2. Obviously, the cardinality has an even value for all iterations except the initial iteration. Therefore, the nth Iteration of Von Koch Curve has 2^n edges then it has 2^{n-1} number of edges that means half of the edges of 2^n are considered as the Matching Matching Cardinality.

7.11.3 KOCH SNOWFLAKE

The Koch snowflake is the most well-known and ancient fractal ever described. This represents the progression of the von Koch curve. Helge Von Koch, the creator

TABLE 7.11
Calculation of Maximum Matching Cardinality of von Koch Curve

n	$E(G) = 2^n$	Count of Vertices $V(G) = 2^n + 1$	von Koch Curve Graph	$M(G) = \frac{E(G)}{2}$
0	1	2		1
2	4	5		2
4	16	17	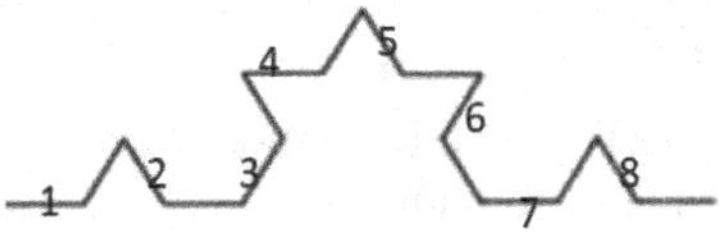	8
6	64	65	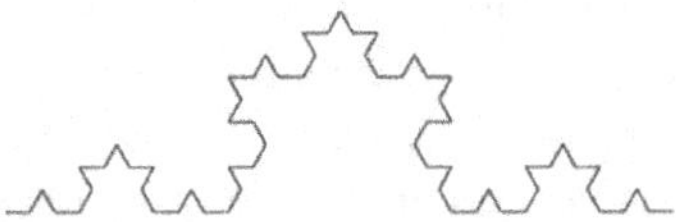	32
8	256	257	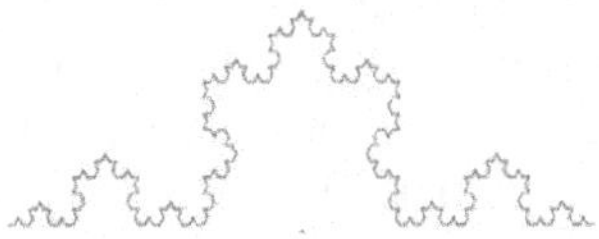	128

of the von Koch curve, also invented the Koch snowflake. Its fractal graph exhibits perfect self-similarity. In the 1870s, Karl Weierstrass proposed the concept of a function that is nowhere differentiable but continuous. In 1906, he refuted the earlier abstract work of Karl Weierstrass regarding the mathematical object of similarity and presented a new study along with an excellent explanation of a continuous curve without a tangent. The fine structure of the Koch snowflake is relatively simple. It begins with an equilateral triangle and can be developed further in Iteration 1. Each segment of the triangle is divided into three parts. The middle section is then moved outward to form a smaller triangle. Each part that is removed is replaced with the new triangle. As the triangles extend outward, a more pronounced resemblance to a snowflake emerges. The sides of the triangle are divided into four segments. This process continues indefinitely, producing a continuous curve that is not differentiable anywhere. The process for the next iteration proceeds in the same manner. The edges of the Koch snowflake increase in common ratio with each iteration. In the nth iteration, the Koch snowflake has $3(4^n)$ edges, where n starts at 0 and is incremented by 1.

7.11.4 Largest Matching in Koch Snowflake

The Koch snowflake is an open-path fractal graph. The largest matching constitutes a collection of non-touching edges. Alternate lines are considered for the maximum matching set, starting at the first edge of the Koch snowflake. The edges of the maximum matching set are denoted as M(G).

Theorem 7.3

Koch Snowflake has $t(4^n)$ edges in n^{th} iterations. Then it has $(2t)\ 4^{n-1}$lines in the Maximum Matching Set for all t is equal to 3 which is an edges of equilateral triangle.

PROOF:

Induction Method is used to prove this theorem.

Step 1:

From the beginning, it is triangle shape. It has only 3 edges. It is given as figure 7.13. Any single edge out of three edges is taken for the Maximum Matching Set. Edges in Koch Snowflake $=3(4^0)=3$. Maximum Matching number $=(2*3)\ 4^0-1=1$ (consider integer only)

Step 2:

It is shown in Figure 7.15, it has 12 edges. Here t=3 and n=1.
Edges of iteration $1=t*4^n=3*4=12$.
Total edges of Maximum Matching Set $=(2t)\ 4^{n-1}=(2*3)*4^{1-1}=6$.
Theorem is true for n =1.

Step 3:

Assuming that it is proven for every n=k, let's proceed. It is made up of of $t\ (4^k)$ edges and $(2t)\ 4^{k-1}$ edges Maximum Matching Set edges.

Step 4:

To prove the statement for all n=k+1. $(k+1)^{th}$ iteration has $t\ 4^{k+1}$ edges. It can be partitioned into two components $(t\ 4^k)$ and 4. In the first partition has $t\ 4^k$ edges of Koch Snowflake and it has $(2t4^{k-1})$ edges of Maximum Matching Set. The second partition has only four edges. It is connected with the first partition by Combinatorics method.

Edges of Largest matching set $2t4^{k-1}*4=2t4^k$. Statement is proved.

7.11.5 Sierpinski Arrowhead Curve through an Iteration Method

The Sierpinski arrowhead curve has an odd count of lines for all iterations. It is an open path structure. The derived formula provided is suitable for this kind of fractal graph, and it follows the common formula for the maximum matching constitutive number. All iterations of the Sierpinski arrowhead curve adhere to the formula: vertices = edges + 1. Here it shows, all the number of edges are powers of 3 (Tables 7.12 and 7.13).

$$\text{Largest matching cardinality } M(G)=\left|\frac{E}{2}\right|+1 \quad \text{or} \quad M(G)=\frac{V}{2} \tag{7.10}$$

TABLE 7.12
Calculation of Maximum Matching Constitutive Number of Sierpinski Arrowhead Curve using an Iteration Method

Iteration	E(G)	$M(G)=\frac{E(G)}{2}+1$	Sierpinski Arrowhead Curve
1	3	2	
2	9	5	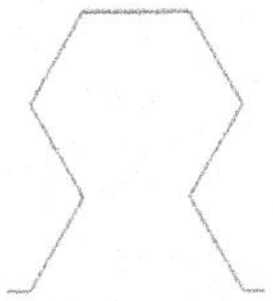
3	27	14	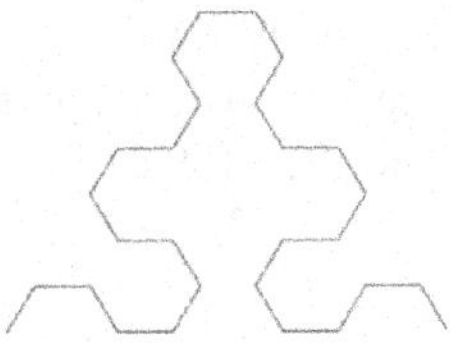
4	81	41	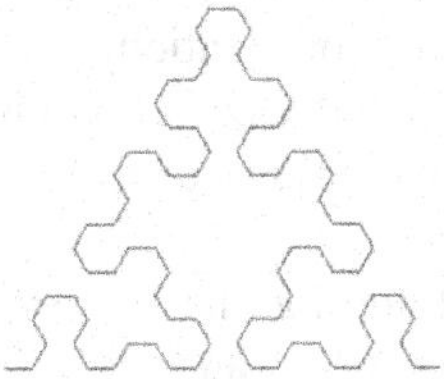
5	243	122	
6	729	365	

TABLE 7.13
Maximum Matching Cardinality in Sierpinski Arrowhead Curve using an Induction Method

n	Graph of Sierpinski Arrowhead Curve	$E(G) = 3^n$	V(G)	$M(G) = \left\lfloor \frac{V(G)}{2} \right\rfloor$ $M(G) = \left\lfloor \frac{E(G)}{2} \right\rfloor + 1$
0		1	2	1
1		3	4	2
2	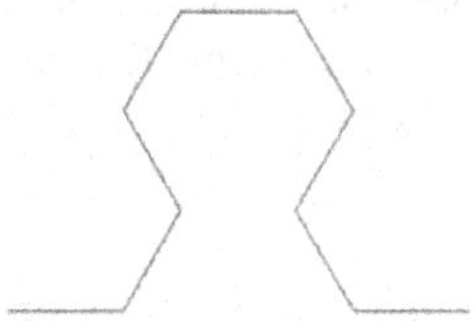	9	10	5
3	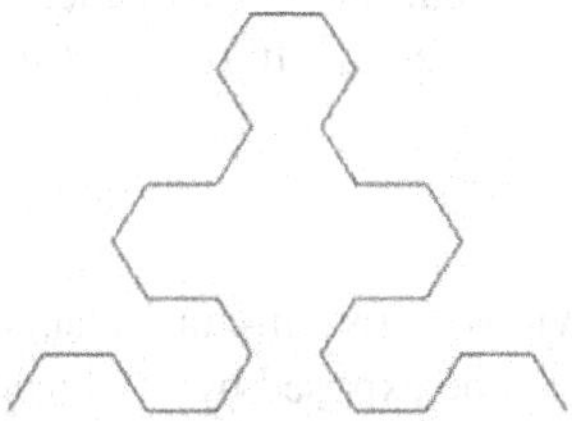	27	28	14
4	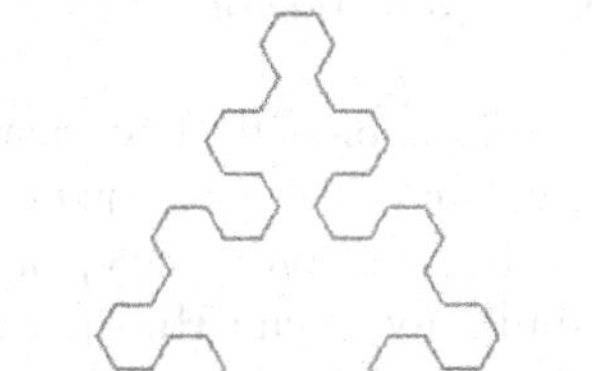	81	82	41
5		243	244	122

7.12 HILBERT FRACTAL CURVE

The Hilbert fractal curve is an important framework for Fractal Antennas and is highly useful compared to other antennas. Fractal design possesses the self-similarity property, which provides enhanced perception, length, resonances, changes, and interpretations. This design disperses electromagnetic radiation throughout the entire surface. Metamaterials, which are man-made objects, are used to construct the Hilbert curve.

7.12.1 Vertices and Edges of Hilbert Curve

7.12.1.1 Iteration 1

The Hilbert curve starts with a unit square missing its top, forming a "U" shape. The bent curve should be straightened to create a square shape (without the top). It can be drawn on a 2×2 cells. The "U" shape begins at the bottom left corner cell; it moves downward and then to the right, followed by an upward movement. The movement of the edges is indicated by arrow

Vertices and Edges of Iteration 1

The Hilbert fractal curve has 4^n vertices, which is a real number that increases by one in each subsequent iteration. There are no crossing edges. In this first iteration, there are three edges and four vertices.

7.12.1.2 Iteration 2

In Iteration 2, every square grid box of the previous iteration can be added into 2×2 grids. The second stage of the Hilbert curve can be exposed by 4×4 grids. The topless square box can be placed in all 2×2 grids. The direction of the open square box is presented into 2×2 grids.

The second stage starts with 2×2 grids. Here, the direction of the first square part is moving rightward. In second 2×2 grids, the direction of the open square part is moving leftward. In third and fourth of 2×2 grids, the direction of the open square part is moving upward. Finally it provides self-similarity fractal Hilbert curve in second iteration. Count of vertices is 16, Count of Edges is 15.

It is shown as in Figure 7.13 arrow mark indicates the direction of movement.

7.12.1.3 Iteration 3

In Iteration 3, 2×2 grids are again enlarged into 4×4 grids, resulting in 8×8 grids. The shapes from the previous iteration are placed in all 4×4 grids, labeled as A, B, C, and D. In the first 4×4 grid shown in the images, the movement is to the right; the second 4×4 grid moves leftward, while the third and fourth grids move upward, as indicated by the arrow marks. This demonstrates how the edges of the Hilbert curve are used to connect the 4×4 grids. In Iteration 3, the total number of vertices is 64 and total number of edges is 63.

Figure 7. 14 depicts the third level of the Hilbert curve.

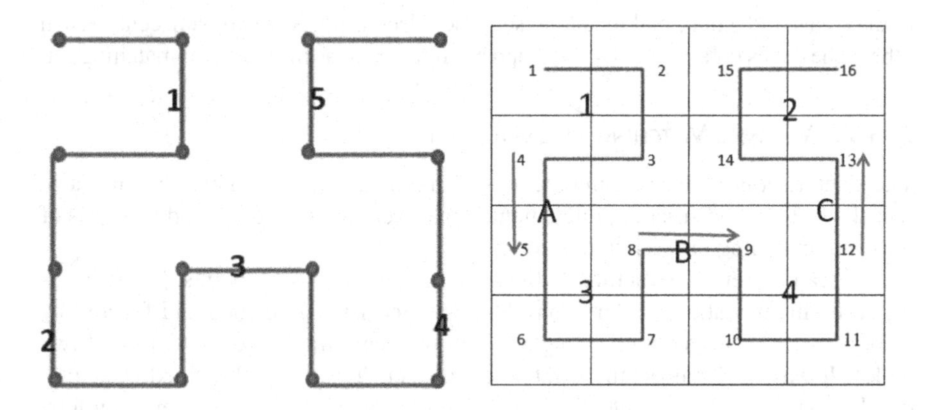

FIGURE 7.13 Second iteration of Hilbert curve

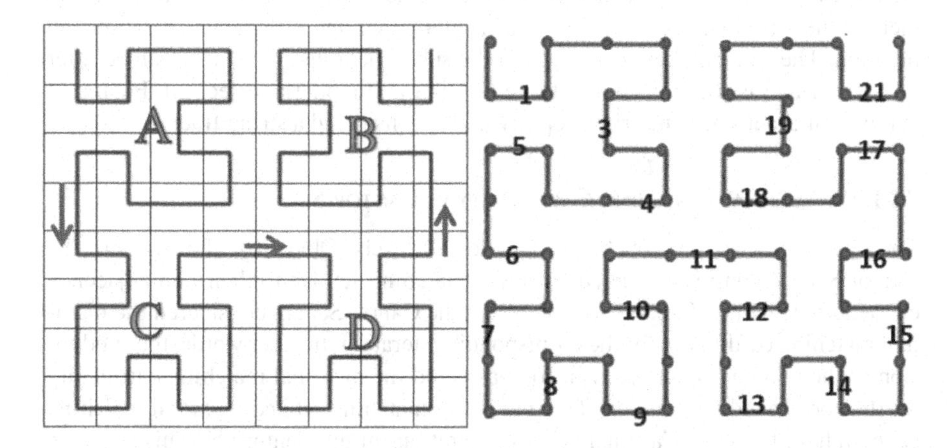

FIGURE.7.14 Third Iteration of Hilbert Curve

7.12.2 Result

Evaluation of Vertices and Edges follows a common ratio. It is implemented in all Iteration. Increase of Vertices incremented by the multiple of four elements. In the first Iteration has 4^1 vertices, second Iteration has 4^2 vertices, third Iteration has 4^3 vertices which implies, in the nth level of Hilbert Curve has 4^n vertices. It is open connected path length. Obviously it is 2 – regular graph. By the property of Graphs, Number of edges of open walk = No. of vertices – 1.

7.13 EXISTENCE OF MAXIMAL MATCHING IN VARIOUS FRACTAL GRAPHS

Maximal matching is a type of matching set. It does not allow even a single line in the matching set. It consists of minimum edges in the matching set. The edges of the maximal matching set are incident to other vertices in the graph that are not incident to the edges of the maximal matching set. The selected edges are non-adjacent to each other. The edges collected from the graph are referred to as the maximal matching set.

7.13.1 Maximal Matching in Cantor Set

The Cantor set is analyzed using the tool of maximal matching. This section examines the history, construction, implementation of vertices and edges, and formulas of maximal matching for each iteration.

The Cantor Set is a structure that has edges passing through points. It was introduced by mathematician Henry John Stephen, and later developed and favored by Georg Cantor. It started with a single line in the beginning stage, consisting of two nodes. It was later named the Cantor ternary set. It has a well-defined structure, developed by a subsidiary rule. Each edge is partitioned into three segments, built by deleting the middle segment. In the first iteration, it has two edges with four vertices. Again, two edges are partitioned into three segments. By deleting the middle segment of the third line, it creates four non-incident lines and eight nodes in the second iteration. The same process continues in the subsequent iterations. It is named after its founder and is recognized as a more familiar and modern perfect set. Each part exhibits an exact self-similarity property and is considered a string fractal.

7.13.2 Maximal Matching Cardinality in Cantor Set

The Cantor Set has non-touching edges for all levels. Obviously, the edges of the Cantor Set are considered the edges of the matching set due to their non-adjacency characteristics. Therefore, the total edges in the Cantor Set are considered the maximal matching cardinality for the corresponding iteration. In other words, the subdivision of the number of vertices is also considered the maximal matching cardinality for the corresponding iteration. The maximum matching set and maximal matching set both have the same Cardinality for all iterations of the Cantor Set. This is shown in Table 7.14.

TABLE 7.14
Maximal Matching Cardinality of Cantor Set

Iteration	V(G)	E(G)	M(G)
1	2	1	1
2	4	2	2
3	8	4	4
4	16	8	8
5	32	16	16
.	.	.	.
.	.	.	.
.	.	.	.
n	2^n	2^{n-1}	2^{n-1}

7.13.3 Maximal Matching of von Koch Curve

The von Koch curve is the oldest prescribed Fractal Graph. In the 19th century, it was developed by Helge von Koch. After that, many mathematicians used the von Koch curve in their papers. It is an exact self-similar fractal graph. The landscape area of Konigsberg can be drawn by vertices, and the bridges connecting the landscape areas are drawn by edges. It consists of four vertices p, q, r, s, and seven bridges. It is the choice of edges that do not intersect. There isn't a single common vertex within this group of edges. The minimal number of non-adjacent edges that are next to the other nodes in the graph is found in the maximal matching set [3]. For the maximal matching set, it is not possible to add a single additional edge to this set. The number of lines in the maximum matching set is known as the maximal matching cardinality. The constant ratio used for each iteration is also analyzed in order to obtain the maximum matching cardinality of the von Koch curve. A fractal graph with an open path is the von Koch curve. The constant ratio between the count of edges and the maximum matching cardinality is implied for every iteration. It is shown as Table 7.15.

TABLE 7.15
Calculation of Maximal Matching Cardinality in von Koch Curve

Iteration	Value of n	Number of Edges $E(G) = 2^n$	von Koch Curve Graph	$M'(G) = \frac{E}{3} + 1$
0	0	1		1
1	2	4		2
2	4	16		6
3	6	64	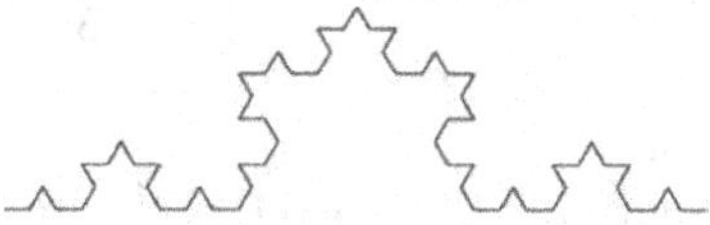	22
4	8	256	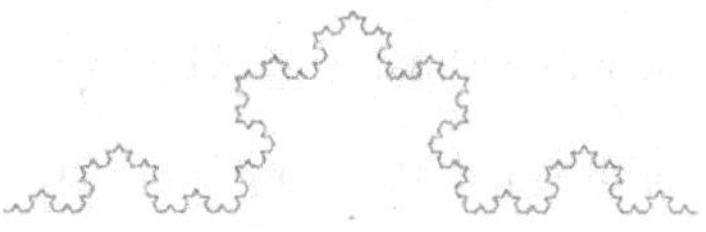	86
5	10	1024	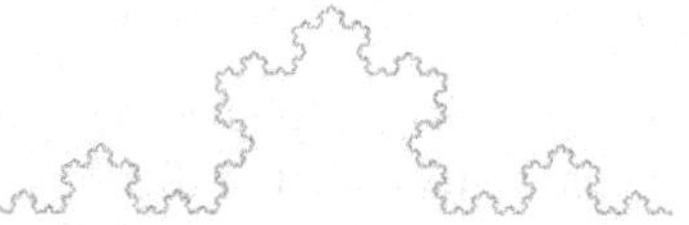	342

Theorem 7.4

Let Odd Open Path Fractal Graph Von Koch Curve has 2^n edges for n^{th} Iteration. Then it has $\frac{E}{3}+1$ edges in Maximal Matching for all n begins at zero and n is incremented by 2.

PROOF:

Maximal Matching Cardinality of Von Koch Curve is evaluated by the Formulae $\frac{E}{3}+1$.

This theorem can be proved and implemented for all Iteration. The largest value of Iteration of the Fractal Graph is not easy to observe. But this theorem is very useful to find edges and cardinality for a large value of Iteration.

Let the Von Koch Fractal Graph has 2^n edges of nth Iteration when is an even number and started at zero and increased by 2.

Let to prove Von Koch Fractal Graph has Maximal Matching Cardinality $M(G)=\frac{E}{3}+1$.

Initial Iteration

Let $n=0$, the number of edges $E(G)=2^0=1$.

Maximal Matching Cardinality $M'(G)=\frac{1}{3}+1 = 0+1=1$. It consists of single edge. It has one Maximal Matching Cardinality. Theorem is proved for Initial Iteration.

First Iteration

Let n=2, n is increased by 2.Number of Edges $E(G)=2^2=4$.

Maximum matching cardinality $M'(G)=\frac{4}{3}+1=1+1=2$. Theorem is proved.

Second Iteration

Here n=4, Number of Edges $E(G)=2^4=16$.

Maximum matching cardinality $M'(G)=\frac{16}{3}+1=5+1=6$.

It has 6 edges Maximal Matching Cardinality.

Theorem is proved for Second Iteration.

In the upcoming level, it also follows the same formula for maximal matching cardinality. It can be tabulated in Table 7.15. The formula continues for all iterations of maximal matching cardinality. Thus, the theorem is proved for all iterations.

7.14 MAXIMAL MATCHING IN HILBERT CURVE

Theorem 7.5 calculates the cardinality of the maximal matching set of the Hilbert curve fractal graph. It is the relevant fine structure of fractal antenna. This section inspects the structure, properties, and increment of vertices and edges, to obtain general formula 7.12 for the evaluation of cardinality in the maximum matching set of the Hilbert fractal graph. This is achieved by the iterative method.

Theorem 7.5

If the Hilbert curve has 4^n vertices and 4^{n-1} edges in the nth iteration, then it has obtained maximal matching cardinality by the formulas $M(G)=\frac{\text{Number of edges}}{3}$.

PROOF:

This theorem is verified by an iterative method. Here n=1. Number of vertices $= 4^n = 4^1 = 4$.

$$\text{Number of edges} = 4^n - 1 = 4^1 - 1 = 3.$$

Any one edge in the graph is selected for Maximal Matching Set.

Maximal matching cardinality $= \frac{\text{Number of edges}}{3} = \frac{3}{3} = 1$.. Theorem is proved.

In the second Iteration, here n = 2. Count of vertices $= 4^n = 4^2 = 16$.

$$\text{Number of edges} = 4^n - 1 = 4^2 - 1 = 15$$

$$\text{Maximal matching cardinality} = \frac{\text{Number of edges}}{3} = \frac{15}{3} = 5$$

Step 3:

Here n = 3. No. of vertices $= 4^n = 4^3 = 64$. No. of edges $= 4^n - 1 = 4^3 - 1 = 63$

$$\text{Maximal matching cardinality} = \frac{\text{Number of edges}}{3} = \frac{63}{3} = 21$$

Result:

$$\text{Maximal matching cardinality} = \frac{\text{Number of edges}}{3} = \frac{4^n - 1}{3} \quad \text{for all } n > 0$$

TABLE 7.16
Maximal Matching Cardinality in Hilbert Curve

Iteration	Figure	No. of Nodes	E(G)	$M'(G) = \frac{E(G)}{3}$
1		4	3	1
2	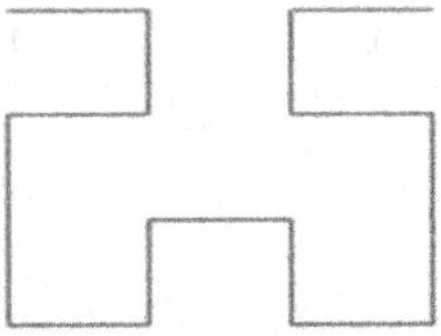	16	15	5
3		64	63	21
4	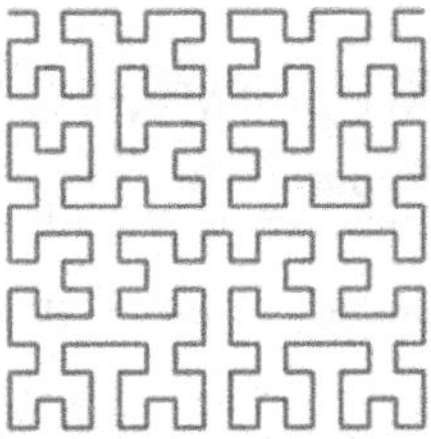	256	255	85
5	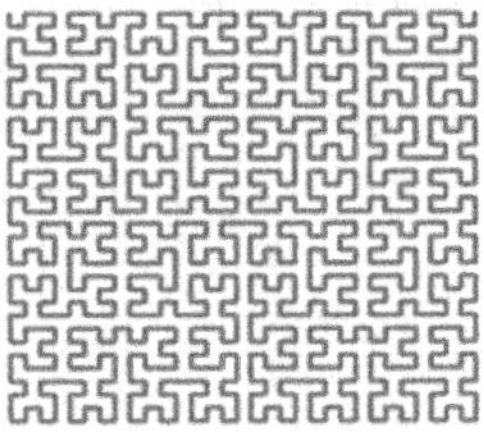	1024	1023	341
6	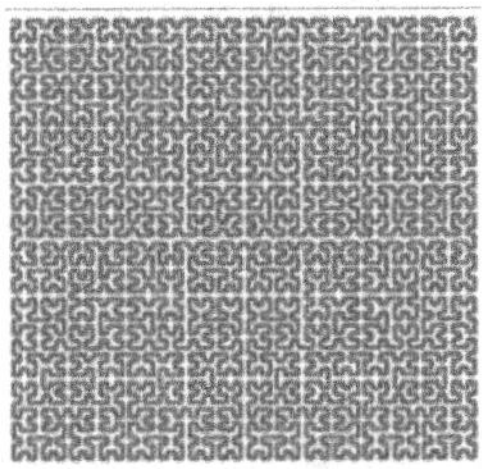	4096	4095	1365

7.15 MAXIMAL MATCHING IN KOCH SNOWFLAKE

It is a closed path and also has an even path length, except in the length of the initial iteration. The Koch snowflake has an odd length path in the initial iteration. The initial iteration starts with an equilateral triangle. Each side can be partitioned into three parts. It consists of an infinite perimeter and finite area. The cardinality of the maximal matching set is evaluated as 1/3 of the edges in the corresponding iteration. Here, one out of every three consecutive adjacent edges can be selected for maximal matching [2].

The maximal matching set considers the minimum level of non-adjacent edges that cover the entire given graph. Here, the initial iteration of the Koch snowflake starts with an equilateral triangle. The maximal matching set has one edge out of three consecutive edges. In the upcoming iteration, each edge can be divided into four segments on all sides. Thus, all iterations of the Koch snowflake must have even path lengths. Maximal matching also has to be increased but in a common ratio. In the initial iteration, we see an equilateral triangle which has an odd path length with three edges. From every three edges, one can be selected for the maximal matching set. This can also be implemented in the upcoming

Theorem 7.6

Let the fractal graph Koch snowflake consist of $E(G)=3*4^n$ edges in the nth iteration. Then, it has $\left(\left|\frac{E}{3}\right|\right)$ edges in the maximal matching set when n is a whole number and is increased by 1.

Note: Here, the modulus attains only whole numbers, and decimal places are negligible.

PROOF:

By using Mathematical Induction Method, Theorem can be proved.

Step 1: In the beginning Iteration n=0. This level has an equilateral triangle. It considered any edge of Maximal Matching. It has one Maximal Matching Cardinality. Number of edges $E(G)=3*4^0=3$. Maximum matching cardinality $M(G)=\left|\frac{3}{3}\right|=1$. Hence is proved.

Step 2: Statement is true for $n=k$ by the assumption.

By the Assumption, in k^{th} Iterations of Koch Snowflake has $3(4^k)$ edges and 4^k Maximal Matching Cardinality.

Step 3: Here to find k+1 Iterations of Von Koch Curve has $3*4^{k+1}$ edges and 4^{k+1} Maximal Matching Cardinality. In this step, Edges can be split into two parts. $3*4^{k+1}$ Edges can be splitted into $3*4^k$ and 3*4 edges.

TABLE 7.17
Maximal Matching Cardinality in Koch Snowflake

S. No.	Value of n	No. of Edges in Iteration I_n	Koch Snowflake Graph	Maximal Matching Cardinality
1	0	3		1
2	1	12		4
3	2	48		16
4	3	192		64
5	4	768		256
6	5	3072		1024

By the assumption, first part of Von Koch Curve has $3 * 4^k$ edges has 4^k Maximal Matching Cardinality. The second part of Koch Snowflake has 12 edges (ie First Iteration) and 4 edges of Maximal Matching Cardinality.

By the method of permutation, the edges of Maximal Matching in two parts can be merged. It implies that $4^k * 4 = 4^{k+1}$ Maximal Matching Cardinality. Thus the theorem is proved.

7.16 MAXIMAL MATCHING IN SIERPINSKI ARROWHEAD CURVE

The open path of the Sierpinski arrowhead curve is an exact self-similar Fractal Graph. It has an interesting structure and is similar to the Sierpinski triangle. The Sierpinski arrowhead curve is a continuous fractal curve that, in the limit case of iterations, becomes a Sierpinski triangle. It can be drawn using an equilateral triangle at equal intervals. It can be developed with two construction rules, turning left and right alternatively at in 60° angles. Here, "+" means it moves left and "–" means it turns right. The Sierpinski arrowhead curve has 3n edges for the nth iteration where n is a real number.

Using the iteration method and induction method, we find maximal matching or minimal matching in the Sierpinski arrowhead curve. It has three edges and four vertices. The middle of the edge was selected for the maximal matching set. The cardinality of the minimal matching is one only. This edge is joined with the remaining edges in the graph. The selected edge for the maximal matching set is colored red. For the minimal matching set, only three non-touching edges have been chosen, which are colored scarlet. All other edges in the graph are adjacent to this one [1].

Only nine non-touching edges have been selected for the minimal matching set. These edges are colored red. This edge touches all the remaining edges in the graph. Continuing the same process, it has 3n vertices and 3n–1 non-adjacent edges in the maximal matching set.

Theorem 7.7

Let the Fractal Graph Sierpenski Arrow Head Curve has 3^n edges in the nth Iteration or level. Then it has 3^{n-1} edges in Maximal Matching Set.

PROOF:

Induction Method is solved this theorem. Let n=1. It has 3^1 numbers of edges. One out of three has selected for Maximal Matching Set. Maximal Matching Cardinality $=31^{-1}=1$. Let n=k. Let us assume that the Fractal Graph Sierpenski Arrow Head Curve has 3^k edges in the kth level. Then it has 3^{k-1} edges in Maximal Matching Cardinality. In the k+1th iteration of Sierpenski Arrow Head Curve has 3^{k+1} number of edges. It has partitioned into two parts 3^k number of edges and 3^1 numbers of edges. By our assumption, 3^k number of edges Sierpenski Arrow Head Curve has 3^{k-1} edges in Maximal Matching Set. Remaining the second partition of three edges has joined by the first partition using Permutation Method. Hence, 3^{k+1} edges occurred for Maximal Matching Set in k+1th level. Hence the theorem is proved

7.17 APPLICATION OF MAXIMAL MATCHING IN FRACTAL GRAPHS

Maximal matching has many applications in various areas such as the construction field, medical field, and design field. These will be explained one by one.

7.17.1 Construction Field

The construction field uses maximal matching cardinality for its work. For example, plumbing works depend on many factors such as pipes, valves, plumbing fixtures, and plumbing adapters. Builders expect that the consumption of plumbing materials wants to be reduced in each construction project. An adapter is one of the main materials used to connect the straight sections of pipes and to fit two dissimilar parts. The cardinality indicates how many adapters are needed for the total plumbing system's connectivity. It is found that some areas have many pipe connections at a single point. Adapters are considered as vertices, and the pipelines are considered as edges. Many edges are incident to a single vertex, which is selected for cardinality. The selected points are only needed for connecting the pipeline adapters. The cardinality impacts the money, time, materials, and tools required for installation at the construction site.

7.17.2 Computer Architecture

In computer architecture, the images selected for cardinality have more connections with other images. In cartoon animation, self-similar images like clouds, trees, and mountains are repeated backgrounds of the pictures. In computer architecture engineering, image processing, and designing the images have played a vital role in real life. Here, the fractal graph is used to set up images to demonstrate screen design, and maximal matching cardinality simplifies image formation

Figure 7.15 shows Background Designing in Self-Similarity Images

7.18 CONCLUSION

Matching has a new growth application. The newly introduced formulae are common for all the corresponding characteristics of graphs. These derived formulae are used to find the cardinality of the maximum and maximal matching for the highest iteration level of the corresponding graph. It is also derived from their number of vertices and edges. The iterative method finds the largest matching cardinality and minimal matching cardinality for the aforementioned graphs. This chapter provides an alternative solution to the famous application of graph theory, such as the Konigsberg

FIGURE 7.15 Background Designing in Self-Similarity Images

Bridge Problem, which is the origin of graph theory. It employs the approach of maximal matching in self-similar fractal graphs. Here, the iterative method is used to find maximal matching cardinality, offering a better alternative to maximum matching. This has numerous useful applications in real life and the engineering field.

It is utilized for computer architecture engineering, image processing, and designing images. It provides the shortest method for finding maximal matching cardinality using the iterative method. Some formulae have been introduced for the evaluation of maximal matching cardinality for certain types of self-similar fractal graphs. The matching tool that locates the non-adjacent edges of the supplied graph is called maximum matching.

This concept is applied across a wide range of fields, including structural engineering, computer architecture, antenna systems, and graphics. This chapter applies the concept of maximum matching to self-similar fractal graphs such as the Cantor set, von Koch curve, Koch snowflake, Sierpinski arrowhead curve, and Hilbert curve. The cardinality of maximum matching has a ratio of vertices to edges for all corresponding iterations. Very large iterations of fractal graphs are too complicated to view and analyze. Thus, the newly introduced formulae are invaluable for evaluating the maximum matching constitutive number at the highest level of fractal graphs.

REFERENCES

1. Anitha, A., Tharaniya, P., Senthil, S., and Jayalalitha, G., Analyzation of maximum matching and maximal matching in various graphs, *International Journal of Pharmaceutical Research*, Vol. 12(3), pp. 778–782 (2020).
2. Gutman, I., Some analytic properties of the independence and matching polynomials, *Match*, Vol. 28, pp. 139–150 (1992).
3. Suguna, R., Tharaniya, P., and Jayalalitha, G., Maximum matching in Sierpenski arrow head curves, *Journal of Advanced Research in Dynamical & Control Systems*, Vol. 12(03–Special Issue), pp. 1380–1385 (2020).
4. Rabin, M. O., and Vazirani, V. V., Maximum matchings in general graphs through randomization, *Journal of Algorithms*, Vol. 10, pp. 557–567 (1989).

8 Domination and Its Applications in Real Life

8.1 INTRODUCTION

An intensively studied subfield of graph theory is dominance in graphs. Graph theory has emerged as one of the most recent developments in computer applications and mathematics, experiencing significant growth and improvement over the past 30 years. Research in this area began around 1950, with Richard Karp establishing the NP-completeness of the set cover problem. The practical applications of this technology include the construction of secure systems for electricity grids, document summarization, and wireless networking, all of which utilize concepts from the dominating set.

Other fundamental dominance problems are also NP-completeness problems and are closely related to other NP-completeness issues. This has played a significant role in the tremendous expansion of the study of domination theory, demonstrating its widespread relevance among graph theory researchers. The careful investigation of dominant sets in graph theory began around 1960, referred to as the "coefficient of external stability." Domination has applications in combinatorial, classical algebraic, and discrete optimization problems. While it has been studied since 1950, the pace of research in dominance sharply increased starting in 1970, following Richard Karp's proof of the NP-completeness of the cover problem. This has influenced the practical approach to the dominating set issue. The current research trend in graph theory indicates its evolution from competitive mathematics into an important area of mathematical investigation [1].

8.1.1 History

The newest trending tool in the theory of graphs is called domination. It plays a significant role in practical applications across various disciplines, including engineering, physical science, and social science. Research on dominating sets began around 1960, with the Journal of Discrete Mathematics publishing "Topics on Domination in Graphs" in its 86th issue in 1990. Following this, domination theory gained substantial popularity within the scientific community. Due to its wide range of applications in fields such as engineering, discrete optimization problems, combinatorial problems, and algebraic problems, it has supported the remarkable expansion of graph theory [2]. It is employed to find efficient routing in wireless networking, design electrical grids, and summarize documents. Although its roots trace back to 1862—addressing the minimum number of queens required to cover or dominate an $n \times n$ chessboard—it continues to be a focal point of research, often referred to as the external stability coefficient.

DOI: 10.1201/9781003481096-8

Hedeniemi presented a detailed and intriguing analysis of dominating sets in graphs in 1977. This paper generated significant interest in the study of domination in graphs. Richard Karp solved the NP-complete set cover problem in 1972, and its implications for the dominating set problem were immediately apparent. The concept of domination is employed in theories of encoding and problems related to facility location. Facility location issues pertain to minimizing the distance required to reach the closest facility. The concept of domination also includes applications in land surveying and communication monitoring. The goal of domination in these contexts is to reduce the number of locations where surveyors take measurements while collecting the highest readings throughout all regions. Since then, the notation D(G) for a graph's dominance number has gained widespread use. The study of domination in graphs has continued to evolve since the publication of Cockayne and Hedetniemi's survey paper.

Over 1,200 research articles on domination sets were published in the roughly 20 years following the survey, with the number of studies continuing to rise. Numerous mathematicians have made important contributions to the field of dominance numbers and related topics. The publication of a book on dominance has provided substantial inspiration, further propelling this field of study. To accommodate the extensive collection of research articles, the area has been categorized into useful subfields. Generally speaking, graph-theoretic parameters differ when employing standard graph-theoretic techniques. Numerous applications exist for dominance, bondage, and domatic numbers in the fields of construction and chemical bonding, among others. These concepts are used to build networks that achieve specified dominance, bondage, or domatic numbers while minimizing the number of vertices and/or edges. Many modified graphs have been constructed based on these fundamental ideas of number theory. Research on the domination issue began in the 1950s, and there was a notable surge in studies on domination in the mid-1970s. The dominant set problem is influenced by the NP-complete set cover problem, as there are straightforward connections between the two problems that relate to vertex bisection [3].

Dominant sets are highly useful in various contexts, contributing significantly to the fields of wireless networking, ad hoc mobile network routing, document summarization, and secure grid design. A subgraph of a graph in which no vertices are adjacent to one another is called a dominant graph. This chapter provides a brief introduction to the principles of dominant sets, covering the determination of dominant sets in different types of graphs, the various kinds of dominating sets, their attributes, trees, augmenting paths, and graph traversal capabilities. Dominating sets are beneficial in numerous applications, including identifying effective paths within mobile networks and constructing secure electrical grid systems as well as document summaries.

8.1.2 Examples of Dominating Sets

The potential collection of the graph's non-adjacent vertices is included in the dominating set D(G). In the illustrated graph, the vertices of the dominating sets are adjacent to each other. The graph with prominent vertices is shown below. The red nodes in the provided graph are selected to form the dominating set, though they do not constitute the incident vertices. In the subsequent graph, these red nodes also connect to the other vertices, represented as white nodes. It has been established that

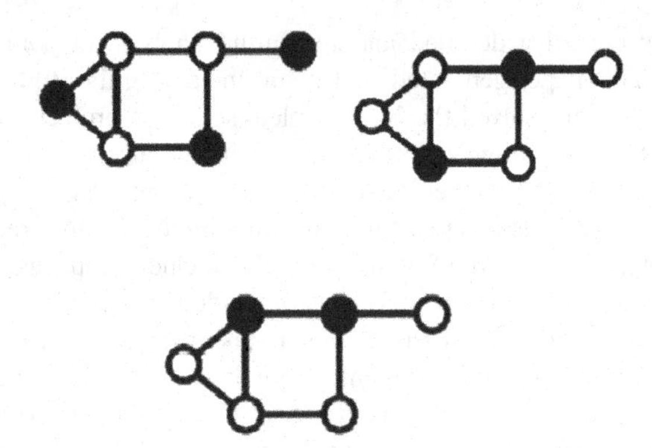

FIGURE 8.1 Maximum dominating vertex: discovered that red nodes dominate white nodes.

red nodes dominate the white nodes. In this example, since every white vertex borders at least one red vertex, the red vertex is said to dominate the white vertex. The dominance number for this graph is 3. Two red nodes have been chosen to form the dominating set, as demonstrated in the earlier examples. Although they are adjacent to other nodes, such as white nodes, these two red nodes are non-touching vertices. It is clear that this graph does not possess a dominating set with just one vertex. As shown in the previous example, the red nodes are not dominating vertices. Therefore, the example above does not represent a dominating graph.

8.2 DOMINATION

The maximum dominating set is the collection of all non-adjacent vertices within the dominating set. It is denoted by D'(G). This set is also referred to as the Maximum Independent Set, represented by D(G). Additionally, the term "domination number" refers to the smallest number of vertices that can be part of the dominating set [4]. The set of vertices that are not adjacent to one another is known as an independent set, where these vertices are not connected by any edges. If no two vertices in a set are neighbors, that set is considered independent. The number of vertices that are not neighboring is called independent cardinality. The domination number represents the total number of vertices in the dominating set and is denoted by D(G) [5]. The term "maximum dominating set" refers to the minimum number of adjacent vertices that can exist, symbolized as D'(G). Additionally, this set is also the maximum independent set.

8.3 DOMINATION IN DIFFERENT TYPES OF GRAPHS

For each graph, click this link to explore the various graph types examined in relation to the concept of domination. The only vertices in a null graph are those that lack edges; thus, the vertices are not all adjacent to one another, as the graph contains no edges. The number of vertices in the graph equals the maximal dominance cardinality and maximum dominance cardinality of the null graph.

$$\text{Maximum domination cardinality} = \text{Maximal domination cardinality}$$

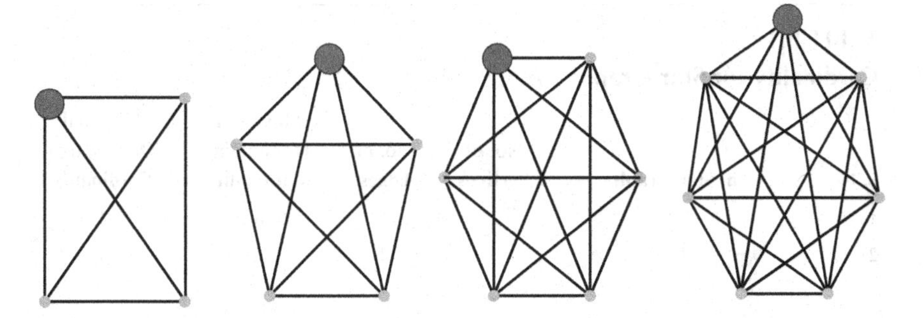

FIGURE 8.2 Value of the maximum domination cardinality and maximal domination cardinality of a complete graph.

8.3.1 Complete Graph

In a complete graph, every vertex is connected to every other vertex via edges. This is a unique type of graph, denoted by the notation Cn, where n represents the number of vertices. The structure is closed, and there is an excess of vertex-to-edge connections. If the graph has n vertices, each vertex has a common degree of $n-1$. The values of maximum domination cardinality and maximal domination cardinality are the same.

Figure 8.2 illustrates the dominating set of the complete graph.

8.3.2 Star Graph

In a star graph, one vertex, known as the center, connects to all other vertices via edges. This type of graph is represented as S(1,n). While the graph appears open, it maintains connectivity through the central vertex. It has n vertices and $n+1$ edges. In this case, the solitary edge is considered the maximal dominating set. The maximum dominating set for this graph includes all vertices except the center vertex. There is one value for maximal domination cardinality in all star graphs. Table 8.1 displays the calculations of maximum domination cardinality and maximal domination cardinality. The generic formula for both maximum and maximal domination cardinality can be found.

8.3.3 Cycle Graph

A cycle graph is both connected and closed. It is denoted as Cn, where $n > 3$. For any cycle graph, the generic or maximum domination cardinality are provided.

$$\text{Maximum domination cardinality} = \left|\frac{\text{Number of vertices}}{2}\right|$$

TABLE 8.1
Cardinality for Star Graph

S. No.	n- Star Graph	No. of Vertices	No. of Edges	Maximum Dominating Cardinality	Maximal Dominating Cardinality
1		2	1	1	1
2		3	2	2	1
3	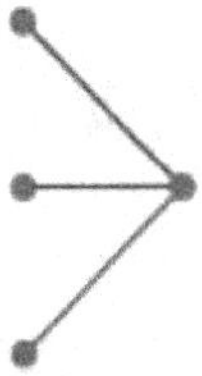	4	3	3	1
4	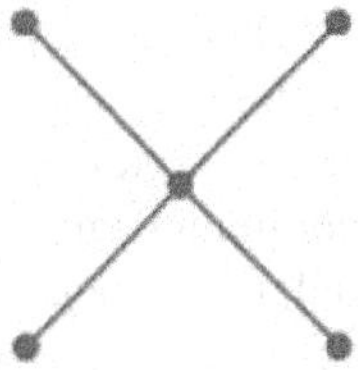	5	4	4	1
5	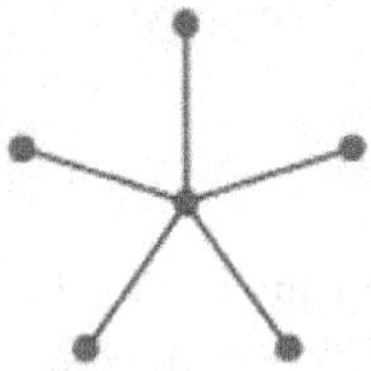	6	5	5	1

$$\text{Maximal domination cardinality} = \left\{ \begin{array}{c} \left| \frac{n}{3} \right| \text{if n is multiple of 3} \\ \\ \left| \frac{n}{3} \right| + 1 \text{otherwise} \end{array} \right\} \quad (8.1)$$

The maximum domination cardinality of the provided graph is determined by its vertex count. Its structure is both closed and looping. To establish a border and a closed loop, all vertices can be connected by an edge to a single vertex or to one another in the correct order. For every wheel graph, the maximal domination cardinality yields the same in the following Table 8.2 and Table 8.3.

TABLE 8.2
Maximal Domination Cardinality for Cycle Graph

S. No.	Cycle Graph	No. of Vertices n	Maximum Domination Cardinality	Maximal Domination Cardinality
1	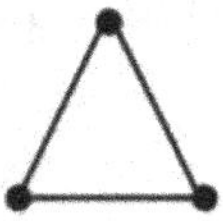	3	1	1
2		4	2	2
3	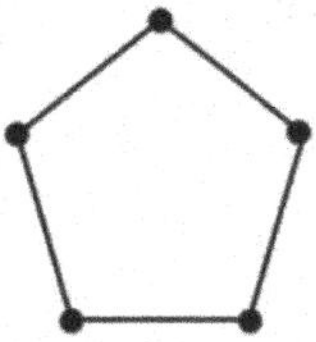	5	2	2
4	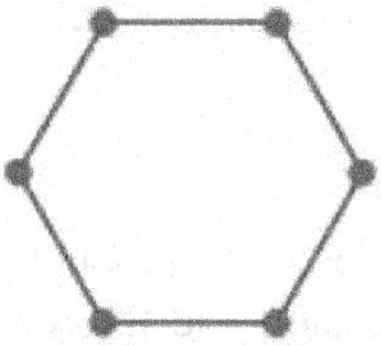	6	3	2
5	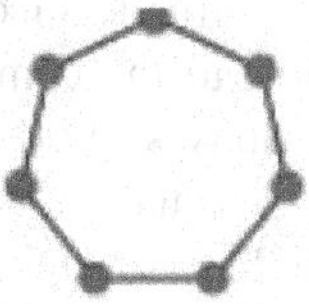	7	3	3

TABLE 8.3
Maximum Dominating Cardinality and Maximal Dominating Cardinality for Wheel Graph

S. No.	Number of Vertices	Wheel Graph	Maximum Domination Cardinality	Maximal Domination Cardinality
1	4	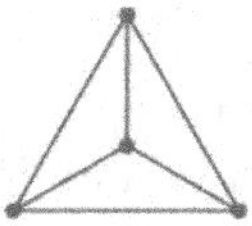	2	2
2	5		2	2

(Continued)

TABLE 8.3 (*Continued*)
Maximum Dominating Cardinality and Maximal Dominating Cardinality for Wheel Graph

S. No.	Number of Vertices	Wheel Graph	Maximum Domination Cardinality	Maximal Domination Cardinality
3	6		3	3
4	7		3	3
5	8		3	3

8.4 DOMINATION IN FRACTAL GRAPHS

Fractals are self-similar graphs that can be broken into smaller pieces, each part's structure visible across the entire shape. Their Hausdorff dimension and finely structured graph are both present [6]. Benoit Mandelbrot, a renowned mathematician, contributed significantly to this field. He discovered a formula applicable to the dominating set shared by all iterations. The implementation of self-similar graphs is based on a shared ratio of vertices, edges, and structures. People are often captivated by the aesthetically pleasing images referred to as fractals. Helge von Koch discovered the von Koch curve in 1904. Later, in 1975, Benoit Mandelbrot presented the concept of fractals to a broader audience. The defining feature of fractals is their self-similarity. They are two-dimensional and more complex than a single dimension, characterized by a simple structure. This structure is defined mathematically and follows a repetitive definition. Generally speaking, objects with smooth or uneven shapes are not characterized by a single set or function. Instead, these objects have been divided into smaller fragments, each exhibiting the same characteristics and mathematical structure

8.4.1 Domination of Cantor Set

The Cantor set is the most fundamental type of fractal graph and is well-known in graph theory. Today, it plays a significant role in various fields, including computer science, animation, data structures, artificial creation, architecture, and more. The Cantor set begins with a single line, which serves as the base level of fractal graphs. Its unit interval is [0, 1]. In the first iteration, a single line is divided into three parts with an interval of 1/3, removing the middle section. The original line was split at

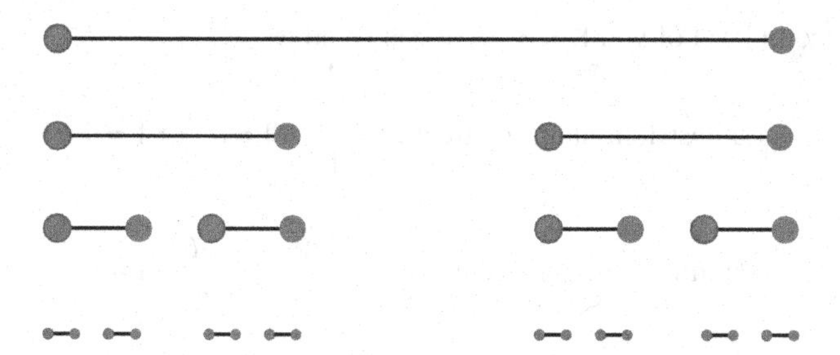

FIGURE 8.3 Value of the maximum domination cardinality and maximal domination cardinality of the Cantor set.

one edge into two segments, and this procedure is repeated for every edge in the graph. Each edge splits into two segments during each iteration. The first iteration begins with a single line containing two vertices. The dominating set consists of just one choice. The product of the maximal and maximum dominating cardinalities is 1.

In the second iteration, the single line splits into two distinct, non-adjacent edges Each edge has one vertex selected for the maximum and maximal dominating sets, resulting in two vertices chosen for the dominating sets.. Again, every edge splits into two halves, and one vertex from each edge is chosen for the dominating sets. In the fourth iteration, there are 16 vertices, and half of these vertices are selected for the dominating sets. There are 2^n nodes and $2^{(n-1)}$ lines in the nth iteration. Two quarters of these vertices, or $2^{(n-1)}$ vertices, are chosen for both maximal and maximum dominating sets.

Figure 8.3 The value of Maximum Domination Cardinality and Maximal Domination Cardinality of Cantor Set

8.4.2 Domination of von Koch Curve

Hedge von Koch created the von Koch curve in 1904. The topological dimension is used first. The fundamental iteration of the von Koch curve begins with the unit interval straight line. This interval is divided into three portions: (0, 1/3), (1/3, 2/3), and (2/3, 1). The center portion is replaced by a two-sided, equal-length triangle with the same diameter as the removed section. . The Von Koch Curve has 2^{2n+1} vertices and 2^{2n} edges (where n is a non-negative integer increased by 1). It are used to compute dominating sets. The first three iterations are covered in the next section. Iterative methods are employed to derive the general formulas for the nth iteration.

Level 1: It has four edges and five vertices.

$$\text{Largest dominating cardinality} = \left|\frac{E}{2}\right| + 1 = \left|\frac{4}{2}\right| + 1 = 3$$

$$\text{Minimal dominating cardinality} = \left|\frac{E}{3}\right| + 1 = \frac{4}{3} + 1 = 2$$

Level 2: Second level of von Koch curve has 16 edges.

$$\text{Largest dominating cardinality} = \left|\frac{E}{2}\right| + 1 = \left|\frac{16}{2}\right| + 1 = 9$$

$$\text{Minimal dominating cardinality} = \frac{E}{3} + 1 = \frac{16}{3} + 1 = 6$$

Level 3: Third level of von Koch has 64 edges.

$$\text{Largest dominating cardinality} = \left|\frac{E}{2}\right| + 1 = \left|\frac{64}{2}\right| + 1 = 33$$

$$\text{Minimal dominating cardinality} = \frac{E}{3} + 1 = \frac{64}{3} + 1 = 22$$

$$\text{Largest dominating cardinality} = \frac{2^{2n}}{2} + 1$$

$$\text{Minimal dominating cardinality} = \frac{2^{2n}}{3} + 1$$

Figure 8.4 illustrates the first iteration of the largest dominating cardinality of the von Koch curve.

Figure 8.5 evaluates the maximum domination set of the first iteration of the von Koch curve.

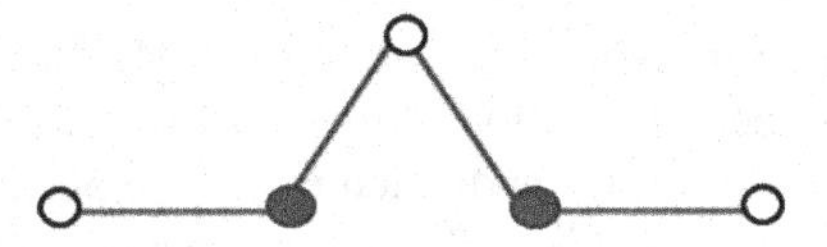

FIGURE 8.4 Cardinality of the maximum dominating set and maximal dominating set of von Koch are evaluated in the first iteration.

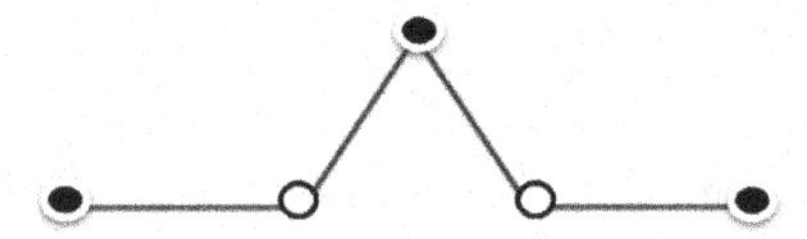

FIGURE 8.5 Cardinality of the maximal dominating set of von Koch are evaluated in the first iteration.

Figure 8.6 represents the second iteration of the maximal domination Set of the von Koch curve.

Figure 8.7 evaluates the maximum domination set of the second iteration of the von Koch curve.

Figure 8.8 depicts the third iteration of the maximal domination set of the von Koch curve.

Figure 8.9 evaluates the maximum domination set of the third iteration of the von Koch curve.

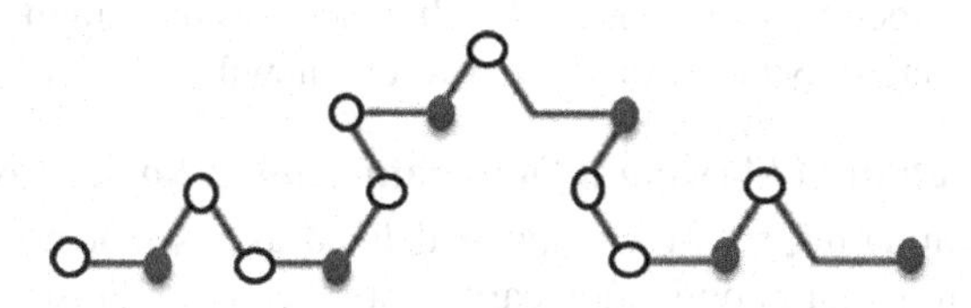

FIGURE 8.6 Cardinality of maximum dominating set of von Koch is evaluated in the second iteration.

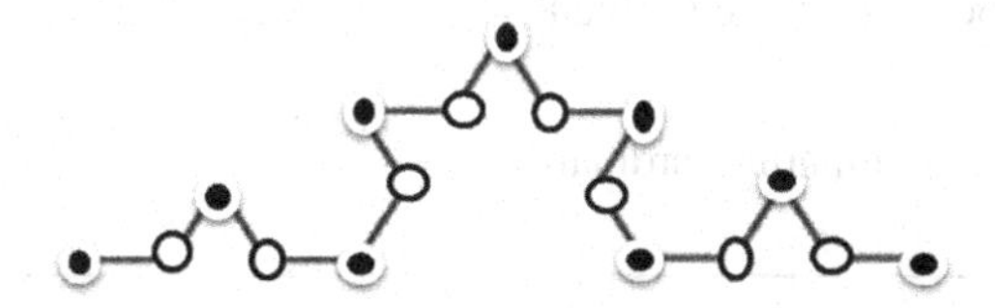

FIGURE 8.7 Cardinality of maximal dominating set of von Koch is evaluated in the second iteration.

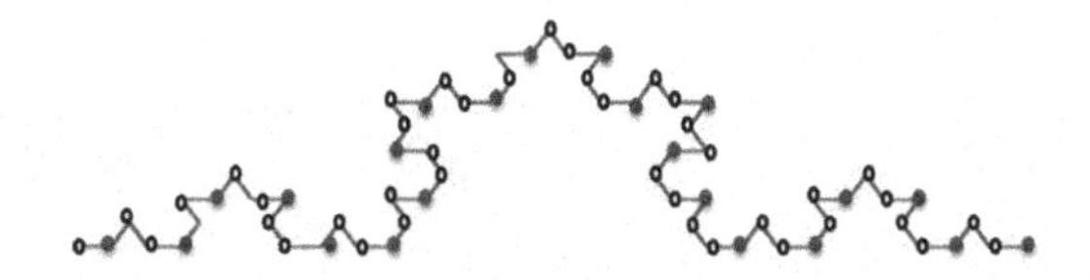

FIGURE 8.8 Cardinality of maximum dominating set of von Koch is evaluated in the third iteration.

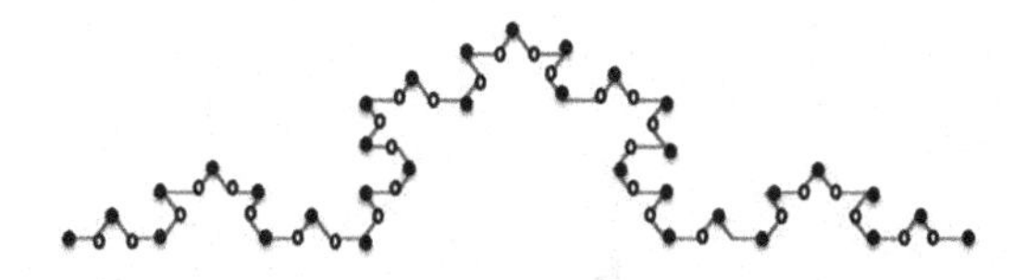

FIGURE 8.9 Cardinality of maximum dominating set of von Koch is evaluated in the third iteration.

8.4.3 Domination of Koch Snowflake

Hedge von Koch invented the Koch snowflake. Karl presented an article which opposed the concept of self-similar objects. His article was disproved by Hedge von Koch, who provided many explanations about geometrical objects that exhibit similarity. The Koch snowflake is constructed iteratively in a sequence of stages. The first level of the Koch snowflake is an equilateral triangle, and each successive level is built by adding outward bends to each side, forming smaller triangles. The total length of the curve increases by a factor of 4/3 with each iteration. The number of line segments in each successive level quadruples compared to the previous level. The perimeter of Koch Snowflake is $3*s*(4/3)^n$ where s is the length of an equilateral triangle. In n^{th} iteration, Number of sides of Koch Snowflake is $3*(4^n)$.

8.4.3.1 Calculation of Maximum Dominating Set in Koch snowflake

A maximum dominating set in a graph is defined as a set of non-adjacent vertices. The term "maximum domination cardinality" refers to the strength of this set, denoted as D(G). In the presented graph, this set of vertices must be in proximity to the remaining vertices that were not selected for the dominating set. It is not always necessary for the selected vertices to be adjacent; thus, alternate vertices are chosen. This chapter computes the maximum and maximal domination cardinality for each iteration of the Koch snowflake (Table 8.4).

$$\text{Maximum dominating cardinality} = \frac{\text{No. of vertices (or) no. of edges}}{2} \tag{8.2}$$

TABLE 8.4
Calculation of Maximum Dominating Cardinality of Koch Snowflake

S. No.	No. of Vertices	Graph of Koch Snowflake	Maximum Domination Cardinality
1	3		1
2	12		6
3	48		24

(Continued)

TABLE 8.4 (*Continued*)
Calculation of Maximum Dominating Cardinality of Koch Snowflake

S. No.	No. of Vertices	Graph of Koch Snowflake	Maximum Domination Cardinality
4	192		96
5	768		384
6	3072		1024

8.4.4 Maximal Dominating Set in Koch Snowflake

It has been established that the Koch snowflake fractal graph possesses a maximum dominating set. A new course on graph theory utilizing fractal graphs has been introduced. Numerous comparable images and objects exist throughout nature. God created natural objects with a structure of similarity that can be classified into four categories. The first category contains objects exhibiting exact self-similarity. Statistically self-similar objects are similar to one another but are not identical. This introduces a combination of two distinct concepts.

The current wave of mathematics involving fractal graphs has closer ties to applications in computer science, such as modeling, cloud computing, imaging, and more. This chapter aims to integrate the two distinct subjects of domination and fractal graphs. The cardinality value in the Koch snowflake fractal graph can be easily evaluated. The cardinality value for each stage of the Koch snowflake is computed using formula 8.3.

$$\text{Maximal dominating cardinality} = \frac{E(G)}{3} = \frac{V(G)}{3} \tag{8.3}$$

8.4.4.1 Iteration 1

The two distinct subjects of fractal graphs and dominance must be combined in this chapter. Determining the cardinality value in the Koch snowflake fractal graph is straightforward. It provides a general formula for computing the cardinality iteratively. This iteration consists of 12 edges and 12 vertices, with three divisions made. The maximal dominating set selects only four vertices, specifically the brown nodes, as shown in Figure 8.10. The decision has been made to use brown nodes for the maximal dominating set, which has been implemented again in Iteration 3. This iteration contains 48 edges and 48 vertices, with one of the three chosen for the maximal dominating set. Only 16 out of the 48 vertices are selected for the maximal dominating set. Table 8.5 provides a tabulation of the maximal dominating cardinality.

TABLE 8.5
Calculation of Maximal Dominating Cardinality of Koch Snowflake

S. No.	No. of Vertices	Graph	Maximal Domination Cardinality
1	3	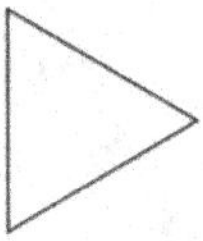	1
2	12		4
3	48	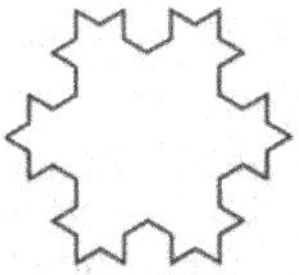	16
4	192	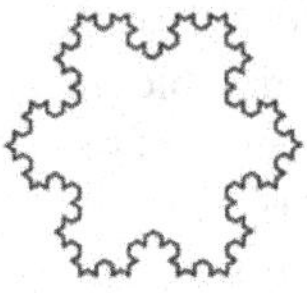	64
5	768	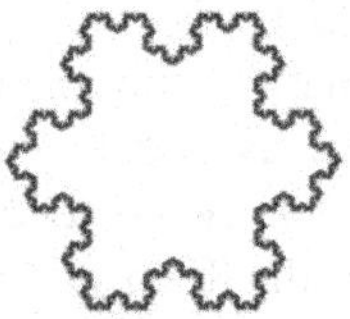	256
6	3,072	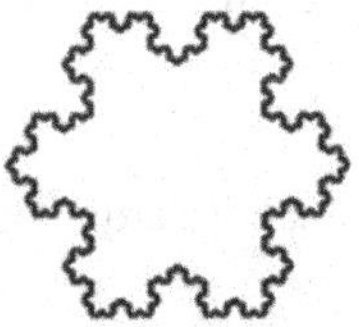	1,024

8.4.5 Maximal Matching in Sierpinski Arrowhead Curve

This is an example of an open path fractal graph, equivalent to the Sierpinski triangle, one of the earliest described and most famous fractal graphs in the world. It is defined by a sequence of continuous, closed plane fractal curves generated recursively. It was introduced by the mathematician Waclaw Sierpinski. This curve is also an example of a space-filling curve and exhibits exact self-similarity. It is used in numerous applications and has more symmetry than other space-filling curves, making it highly regarded. The construction begins with a half-hexagon shape. In the subsequent First Iteration, each edge is adjusted, with horizontal and vertical lines turning into a 60° angle, moving to the left and right. The vertical line moves in the pattern *H-R-V-R-H,* while the horizontal line follows *V-L-H-L-V* where L for left, H for horizontal, and R for Right. The terminal vertex remains fixed. Each edge increases in multiples of three, and since the Sierpinski curve is a space-filling curve, as n approaches infinity, the n-th iteration curve S_n's Euclidean length is

$$l_n = \frac{2}{3}\left(1+\sqrt{2}\right)2^n - \frac{1}{3}\left(2-\sqrt{2}\right)\frac{1}{2^n} \tag{8.4}$$

8.4.5.1 Evaluation of Vertices and Edges

In the first iteration, it has a half-hexagonal shape with three edges and four vertices. Each edge has three multiples. The Sierpinski arrowhead curve comprises ten vertices and nine edges in its second iteration, 28 vertices and 27 edges in its third iteration, and 82 vertices and 81 edges in its fourth iteration. Table 7.8 contains these calculations.. Typically, the Sierpinski arrowhead curve $3^n + 1$ vertices and 3^n edges in its nth iteration.

$$\text{Maximum domination cardinality} = D(G) = \frac{V}{2} \tag{8.5}$$

$$\text{Maximum domination cardinality} = D'(G) = \frac{E}{3} + 1 \tag{8.6}$$

TABLE 8.6
Maximal Dominating Cardinality in Sierpinski Arrowhead curve

Iteration	Nodes V	$D=\frac{V}{2}$	E	$D'=\frac{E}{3}+1$	Sierpinski Arrowhead curve
1	4	2	3	2	

(Continued)

TABLE 8.6 (*Continued*)
Maximal Dominating Cardinality in Sierpinski Arrowhead curve

Iteration	Nodes V	$D=\frac{V}{2}$	E	$D'=\frac{E}{3}+1$	Sierpinski Arrowhead curve
2	10	5	9	5	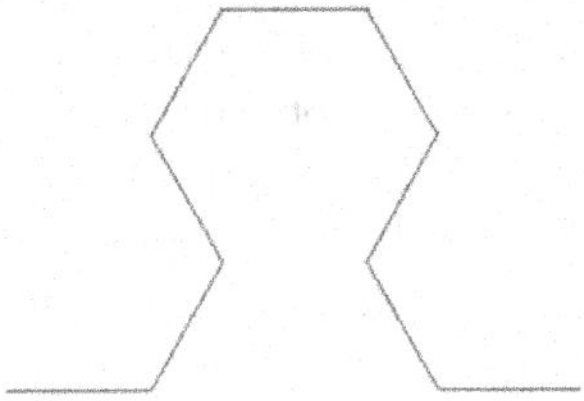
3	28	14	27	14	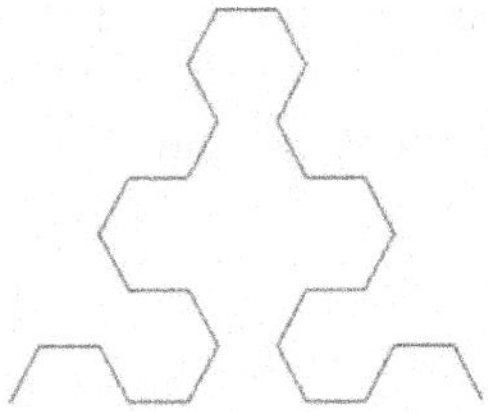
4	82	41	81	41	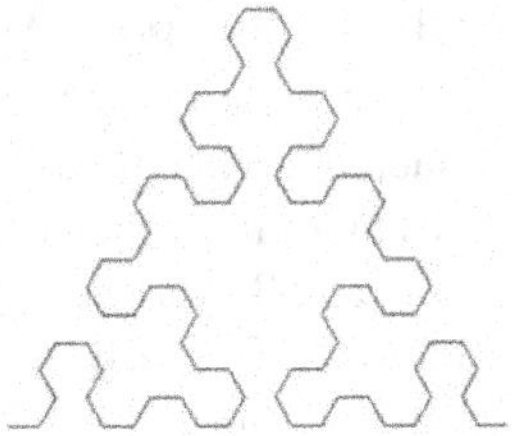
5	244	122	243	122	
6	730	365	729	365	

8.5 APPLICATION OF DOMINATION SETS IN REAL LIFE

One of the fundamental concepts in the real world is the concept of domination. Numerous examples illustrate this, including identifying the neural pathways that connect most of the body's organs, positioning a camera to maximize its field of view, and selecting locations for stone ornaments to enhance their beauty. Analyzing the application of domination is essential for further projects.

8.5.1 Headlamps with Domination

Bicycle lighting, combined with reflection, enhances visibility for riders. In conditions of low ambient light, it helps riders locate other vehicles. Bicycle lighting is primarily used to improve road visibility, allowing cyclists to see their surroundings. Proper placement of bicycle lights requires more power and light output. Headlamps and bicycle lights are typically situated near each other; they serve to indicate the direction of the bike. As lighting conditions deteriorate, particularly on poorly lit roads, the significance of bicycle lights increases. The advantage of lighting for cyclists is that it enhances visibility. Roadside directional indicators are often positioned at higher elevations. Different types of bicycles have various headlamp mounting locations. The main purpose of headlamps on bicycles is to increase illumination and reduce darkness while riding. The front handlebars are the most dominant feature of the bicycle. This part is superior to other bicycle components in terms of effectiveness.

8.5.2 Spinal Card with Domination

The nerve tissue that makes up the human body's long, thin, fibula-shaped spinal cord is known as the spinal cord. The medulla oblongata is a key structure of the brain stem. The section of the vertebral column that houses the medulla oblongata allows the spinal cord to expand. Cerebrospinal fluid is present in its central canal. The human brain and spinal cord monitor the activities of the central nervous system; consequently, the spinal cord is the most dominant site of the neurological system in the human body. One of the constantly growing fields of mathematics is graph theory. Its expansion can be attributed to its applications in a wide range of domains, including engineering, linguistics, discrete optimization, combinatorial science, physical sciences, social sciences, biological sciences, and classical algebraic issues. This thesis focuses on two main topics: domination theory in graphs and various domination parameters in graph-valued functions.

Many mathematicians have revisited graph theory multiple times to address problems pertinent to their areas of specialization. Numerous subfields in graph theory, such as graph coloring, graph decomposition, graph labeling, and dominance, have emerged from these problems. The primary focus of this thesis is the application of domination theory to graphs. The study of dominance in graph theory is a rapidly growing area. While the concept of domination is not new, with early discussions appearing in the writings of Ore and Berge, the literature summarizes several well-known problems considered among the earliest applications of dominating sets. For example, an army post is situated in a designated area to share borders with the

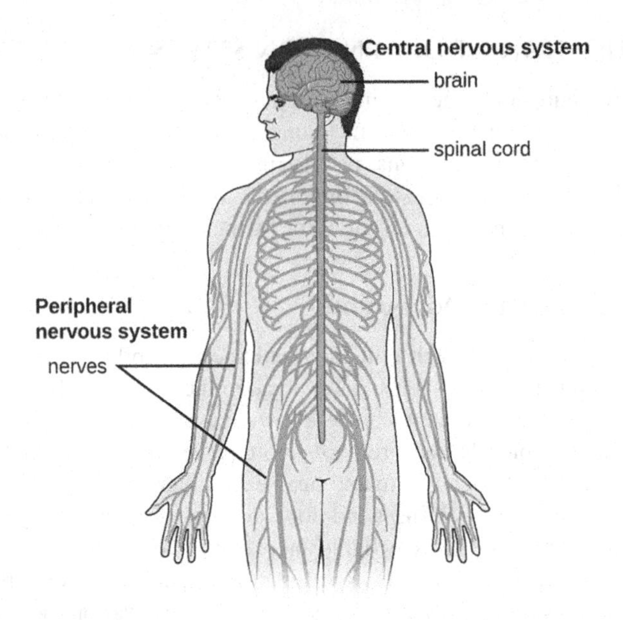

FIGURE 8.10 Domination set of the central nervous system and peripheral nervous system

surrounding region. The objective is to determine the fewest number of post offices and their locations that can adequately service the entire area. We model this scenario using a graph in which each small area is represented by a vertex, and edges connecting two vertices represent areas that share a common border.

Figure 8.10 shows Domination Set of central nervous system and peripheral nervous system

8.5.3 School Bus Routing Problem

This is a standard regulation that most school districts adhere to: every child within a designated region must be reachable by every bus route arranged within a quarter of a mile. Regulations may also stipulate that no bus travel should exceed a specific number of minutes.

8.5.4 Radio Stations

There are restrictions on the transmission range of radio stations. To ensure coverage for every village, we must establish multiple stations. Due to the high costs involved, we use dominance in graphs to determine the minimum number of stations needed, thereby benefiting all the villagers in the process.

8.5.5 Multiple Domination Problems

Numerous dominances have a significant effect. Multiple dominances can be utilized to create hierarchical overlay networks in peer-to-peer applications, which can speed up index searches. Hierarchical overlay networks are commonly used in modern computer applications, such as file sharing and instant messaging. Both the distributed construction of minimum spanning trees and the trade-off between fault tolerance and efficiency utilize multiple types of dominating sets.

8.5.6 Wireless Sensor Networks

Wireless sensor networks consist of spatially distributed autonomous sensor nodes that track physical or environmental conditions, and they have a wide range of applications in both human communities and the larger world. Like other computer and telecommunication networks, wireless sensor networks are susceptible to devastating threats and attacks. Network defense techniques cannot be applied to these electronic devices due to their minimal hardware. One of the main coding tasks in all types of applications where security is an issue is key placement. Despite the limited resources on sensor nodes, it is possible to use existing routing and security procedures in wireless sensor networks. A secure routing protocol design is essential for wireless sensors due to the effects of physical sensor nodes and resource limitations.

8.6 CONCLUSION

This chapter explains the features of different graphs and how vertices and edges are implemented. The relationship between the vertices and edges, as well as the cardinality of the dominating set, is explored. General formulas can also be applied for implementation. Future work needs to address implementation into other types of complex graphs, such as Petersen graphs and fractal graphs, where the dominating set has many applications in real-life scenarios. The basic formulas for nodes and edges in various graphs, particularly fractal graphs, are evaluated in this chapter. For the graphs discussed above, the domination cardinality of the highest iteration is assessed. The cardinality point also serves as the location of the camera on the cars, indicating the camera's maximum projection and focusing on its coverage area. Another example of a dominating set in a room is the placement of light fixtures; increased illumination enhances visibility, making full space coverage the primary task. Consequently, the dominating point is the carefully chosen location that provides the greatest projection.

REFERENCES

1. Bollobas, B., and Cockayne, E. J., Graph Theoretic Parameters concerning domination, independence and irredundance, *Journal of Graph Theory*, Vol. 3(3), pp. 241–249 (1979).
2. Bondy, J. A., and Murty, U. S. R., *Graph Theory with Applications*, North-Holland, New York (1976).
3. Bresar, B., Henning, M. A., and Rall, D. F., Paired-domination of Cartesian products of graphs and rainbow domination, *Electronic Notes in Discrete Mathematics*, Vol. 22, pp. 233–237 (2005).
4. Fomin, F. V., and Dimitrios, M., Dominating sets in planar graphs: Branch-width and exponential speed-up, *Society for Industrial and Applied Mathematics Journal on Computing*, Vol. 36(2), pp. 281–309 (2006).
5. Cockayne, E. J., Gamble, B., and Bruce, S., An upper bound for the k-domination number of a graph, *Journal of Graph Theory*, Vol. 9(4), pp. 533–534 (1985).
6. Bange, D. W., Barkauskas, A. E., Host, L. H., and Slater, P.J., Efficient near domination of grid graphs, *Congressus Numerantium*, Vol. 58, pp. 83–92 (1987).

9 Coloring and Its Applications in Real Life

9.1 INTRODUCTION

Coloring the vertices involves assigning a color to each one, ensuring that adjacent vertices have different colors. Each edge's terminal vertices must have distinct hues, such as red, yellow, blue, green, etc. The chromatic number (CN) represents the minimum number of colors required to color the vertices of the graph. This chapter also analyzes the dominated CN of various graphs, including the Sierpinski arrowhead curve, Hilbert curve, von Koch curve, Koch snowflake, complete graph, star graph, wheel graph, Cantor set, and trivial graph. It was discovered that these graphs require a constant number of colors for proper coloring. This is a well-developed concept in the research area. The chapter further explores the CN in various graphs, particularly focusing on fractal graphs exhibiting self-similarity.

Figure 9.1 provides an example of coloring.

9.2 GROWTH OF COLORING IN THE RESEARCH AREA

The study of the four-color problem began in 1852 when Augustus de Morgan wrote a letter to his friend William Hamilton. In this letter, he inquired whether it was possible to color the regions of a map using just four colors so that neighboring regions had different colors [1]. This question later became known as the four-color problem. The problem was initially posed by Francis Guthrie, who established in his paper that only four colors were necessary for coloring the regions of an English map while ensuring that neighboring countries were assigned different colors. Guthrie handed the problem to his brother Frederick, who was an undergraduate mathematics student at Cambridge, asking if this was the case for every map. It was proved that five colors are always sufficient; however, despite extensive efforts, a generally accepted solution to the four-color problem was not published until 1977. We will consider two types of graph coloring in this context [2].

9.3 LABELING THEORY

Without further explanation, a vertex-labeled graph with distinct labels is referred to as a "labeled graph." The consecutive integers $\{1, 2, 3, \ldots, |V|\}$, where $|V|$ is the

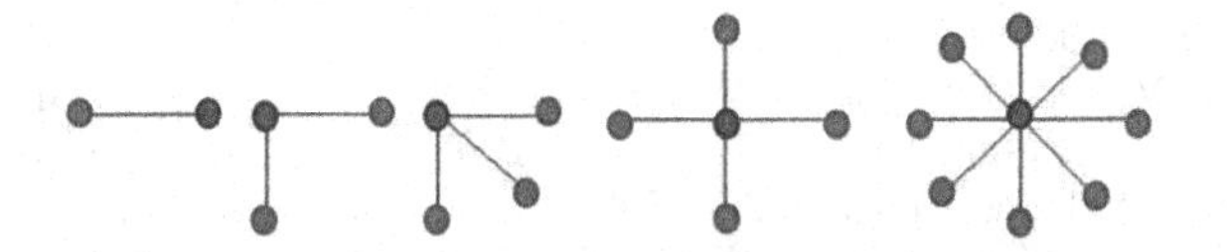

FIGURE 9.1 Example of coloring in different iterations of star graph.

DOI: 10.1201/9781003481096-9

number of vertices in the graph, can also be used to label such a graph. In numerous applications, meaningful labels from the corresponding domain are assigned to the edges or vertices. For instance, the cost of traveling between incident vertices could be represented by weights assigned to the edges. According to the above definition, a graph is a finite, undirected, simple graph. However, the concept of labeling can also be extended to all graph extensions and generalizations. For example, labeled multigraphs—where a pair of vertices is connected by multiple labeled edges—are useful in automata theory and formal language theory. The majority of graph labeling can be traced back to the labeling introduced by Alexander Rosa in his 1967 paper. Rosa classified labeling into three categories: α-, β-, and ρ-labelings. Solomon Golomb later called β-labelings "graceful", and the term has remained in use since.

In graph theory, one specific application of graph labeling is called graph coloring. This involves assigning labels—traditionally called "colors"—to parts of a graph while considering specific limitations. The process of coloring a graph's vertices so that no two neighboring vertices share the same color is known as "vertex coloring" in its most basic form. Face coloring on a planar graph works similarly, assigning a color to each face or region such that no two faces that share a border have the same color. Similarly, edge coloring colors every edge so that no two neighboring edges have the same hue. A graph's coloring is almost always a suitable vertex coloring when done unconditionally; that is, coloring the graph's vertices in a way that ensures no two vertices sharing an edge have the same color. Since a vertex with a loop, or a connection back to itself, could never be colored appropriately, graphs in this context are said to be loopless. The term "coloring maps" refers to the process of assigning colors to vertex labels. When there are not many colors available, only names like red and blue are used, and it is usually thought that these labels originate from the integers {1, 2, 3, …}.

9.4 VERTEX COLORING

One idea in the mathematical field of graph theory is vertex coloring. Simply put, it involves assigning different colors to vertices (points or nodes) in a network so that no two neighboring vertices (connected by an edge) share the same color. Since other coloring problems can be converted into instances of vertex coloring, it is commonly used to introduce graph coloring problems. The vertex coloring problem entails calculating the CN and a valid coloring for a graph. Vertex coloring is applicable in many areas, including computer science, operations research, scheduling difficulties, map coloring, register allocation in compilers, and numerous optimization problems.

9.5 EDGE COLORING

One can infer the edge coloring of a graph from the vertex coloring of its line graph, and the face coloring of a plane graph from the vertex coloring of its dual. This is partly because some problems, such as edge coloring, are better studied in their non-vertex form, and partly because it serves as an instructional exercise.

9.6 GRAPH COLORING

The initial findings on graph coloring, which took the form of colored maps, primarily dealt with planar graphs. It was noted that four colors would suffice to ensure that no two adjacent areas shared the same color. In a paper published that year, Alfred Kempe claimed to have proven this assertion, leading to a ten-year belief that the four-color problem had been solved. Kempe's achievements earned him a fellowship in the Royal Society, and he later became the president of the London Mathematical Society.

9.7 GROWTH OF COLORING IN RESEARCH AREA

In 1890, Heawood argued that Kempe's reasoning was flawed. However, in that work, he proved the five-color theorem, which states that a planar map can be colored with only five colors, by utilizing Kempe's theories. Numerous studies and theoretical efforts throughout the following century sought to limit the number of colors to four, culminating in Kenneth Appel and Wolfgang Haken's proof of the four-color theorem in 1976.

George David Birkhoff first introduced the chromatic polynomial in 1912 to investigate coloring difficulties. Tutte later expanded it to create the Tutte polynomial, which has become one of the fundamental structures in algebraic graph theory [3]. Building on Kempe's discovery of the general, non-planar situation in 1879, a number of results regarding the application of planar graph coloring to higher-order surfaces were published in the early 1900s. Graph coloring has been studied as an algorithmic problem since the early 1970s; Karp published 21 NP-complete problems in 1972, including the CN problem. Concurrently, several exponential time algorithms based on backtracking and Zykov's deletion–contraction recurrence (1949) were developed. One of the primary applications of graph coloring, first seen in compilers in 1981, is register allocation.

9.8 CHROMATIC POLYNOMIAL

The chromatic polynomial measures how many different color combinations are feasible within a given range for coloring a graph. For example, the graph in the following figure can be colored in twelve different ways using just three colors. Two colors are completely insufficient for coloring it. With four colors, it can be colored in $24+4 * 12=72$ ways: there are $4!=24$ legitimate colorings (any four-color assignment to any four-vertex graph is a proper coloring); (Table 9.1).

Available colors: 1, 2, 3, 4, …

Number of Colorings: 0, 0, 12, 72, …

As the name suggests, the function is in fact a polynomial in t for a given G. For the example graph, $P(G, t) = t\,(t-1)^2\,(t-2)$ and indeed $P(G, t) = 72$.

9.9 APPLICATIONS OF GRAPH COLORING

The goal of the graph coloring problem is to assign specific colors to parts of a graph while adhering to predetermined guidelines. Here are a few issues that the concepts behind graph coloring techniques can help address.

TABLE 9.1
Chromatic Polynomials for Certain Graphs

Triangle	$t\,(t-1)(t-2)$
Complete graph	$t\,(t-1)(t-2)\ldots(t-(n-1))$
Tree with n vertices	$t\,(t-1)^{n-1}$
Cycle	$(t-1)^n+(-1)^n(t-1)$
Petersen graph	$t(t-1)(t-2)(t^7-12t^6+67t^5-230t^4+529t^3-814t^2+775t-352)$

5				7				3
							3	
		8		2				
						9		
				0				
		3		0				
		5						
		7						9

FIGURE 9.2 Sudoku.

9.9.1 Sudoku

A variation of the graph coloring problem, Sudoku is one of the most interesting number placement puzzles. Each cell represents a vertex, and an edge connects two vertices that are in the same block, row, or column.

Figure 9.2 is an example of Sudoku.

9.9.2 Register Allocation

One method of compiler optimization is register allocation, which stores the values used most frequently in the fast processor registers. This technique improves the execution time of the resulting code. Registers should ideally have values assigned to them so that when they are used, all of the necessary values can reside there. The problem is modeled as a graph coloring problem, according to the thesis approach. When two vertices are required simultaneously, an edge connects them in the interference graph that the compiler creates. If k colors can be used to color the graph, then any combination of variables required at the same time is achievable.

9.9.3 Scheduling

Many scheduling issues are addressed by vertex coloring models. In its most straightforward form, time slots must be assigned to a specific set of jobs, with each job requiring a single slot. Tasks can be arranged in any sequence, but if two jobs depend on the same resource, a conflict may arise between them. Each job corresponds to a vertex in the graph, and every pair of conflicting jobs has an edge connecting them. The CN of the graph represents the smallest makespan, or the ideal amount of time needed to complete all tasks without encountering any conflicts. The structure of the graph is defined by the specifics of the scheduling issue. For instance, the coloring problem can be effectively solved when assigning aircraft to flights, resulting in an interval graph as the conflict graph. The conflict graph that emerges from allocating bandwidth to radio stations is a unit disk graph, indicating that the coloring problem is three-approximable.

9.9.4 Job Scheduling

In this scenario, the jobs are represented as the graph's vertices, with two jobs connected by an edge to indicate that they cannot be completed concurrently. The coloring of the graph and the feasible scheduling of the jobs have a one-to-one correspondence.

9.9.5 Airline Scheduling

The terms "planning," "scheduling," and "assignment" refer to the processes involved in managing aircraft flights within an airline's operations. This encompasses considering variables such as aircraft availability, crew schedules, passenger demand, and operational efficiency. The goal is to determine which aircraft will fly which flights, when, and on which routes. Assume that n flights must be assigned to each of the k planes, with each flight occurring in the interval (ai, bi). The same aircraft cannot be assigned to two overlapping flights. The vertices of the network represent the flights, and if their time periods overlap, the corresponding vertices will be connected. As a result, the graph forms an interval graph, enabling optimal coloring to be achieved in polynomial time.

9.9.6 Time Table Scheduling

When the constraints are complex, assigning classes and subjects to professors becomes a primary challenge. Graph theory plays a crucial role in solving this issue. Teachers instructing n subjects must develop a schedule for the available p periods. This is carried out as described below.

The vertices of a bipartite graph G represent the number of professors (m1, m2,..., mk) and subjects (n1, n2,..., nm). Edges connect the vertices. It is assumed that each professor can instruct a maximum of one subject at a time, and that a subject can be taught by a maximum of one professor. Consider the first era as an illustration. Therefore, the solution to the timetabling problem can be found by dividing the edges of graph G into the fewest feasible matchings. The edges must also be colored with the minimum number of colors.

Figure 9.3 illustrates an example of the four-color problem.

Figure 9.4 illustrates an example of the coloring problem

9.9.7 GSM Phone Networks

GSM is a mobile phone network that is geographically divided into hexagon-shaped cells or regions. Every mobile phone searches for neighboring cells to connect to the GSM network. Graph theory indicates that only four colors can be used to color the cellular regions because GSM operates in four different frequency ranges. This means that these four distinct colors are necessary for proper coloring of the regions. Consequently, any GSM mobile phone network can utilize the vertex coloring algorithm to assign a maximum of four distinct frequencies.

9.10 APPLICATION OF COLORING IN REAL LIFE

Graph coloring has many applications in our daily lives. It simplifies complex puzzles and critical problems. Graph coloring is becoming an advanced research tool for the coming years. Some of the applications are explained in the following sections.

9.10.1 Applications Map of the Region

Two states that share a boundary are represented by distinct colors on our national map. A graph can be used to illustrate this concept. Each state in the nation is

P	n_1	n_2	n_3	n_4	n_5
m_1	2	0	1	1	0
m_2	0	1	0	1	0
m_3	0	1	1	1	0
m_4	0	0	1	1	1

FIGURE 9.3 Four-color problem.

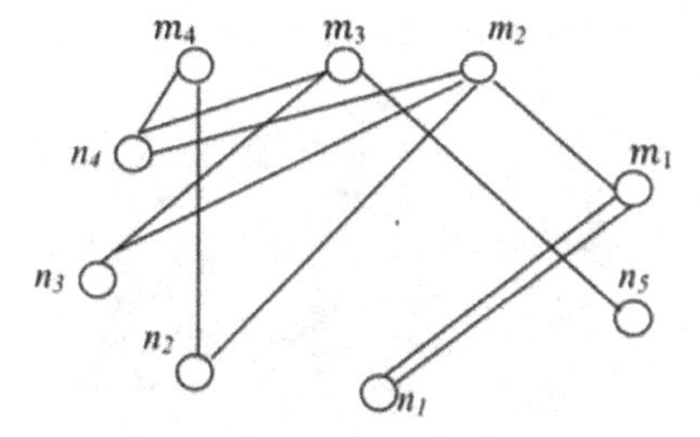

FIGURE 9.4 Example of a coloring problem.

symbolized by a vertex, and edges represent common borders between two states. The resulting graph is referred to as the dual of the country map.

Figure 9.5 provides an example of coloring in the India map.

9.10.2 GSM Network

The Mobile Telephone System, known as Group Special Mobile (GSM), was introduced in 1982. The first GSM network was launched in Finland in 1991 and provides joint technical infrastructure maintenance from Ericsson. It is a world-renowned network used by billions of people in over 300 countries. The GSM network utilizes only four different frequency ranges. All mobile phones connect to the GSM network through nearby cell towers. At the research level, the cellular regions can be effectively colored using just four different colors. This demonstrates that four distinct frequencies are sufficient to cover all the regions of the given map.

Figure 9.6 illustrates the GSM mobile phone network.

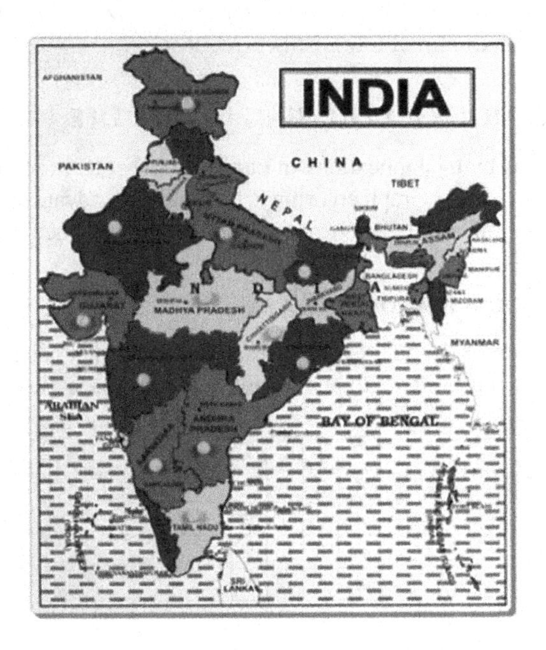

FIGURE 9.5 Coloring in Indian National map.

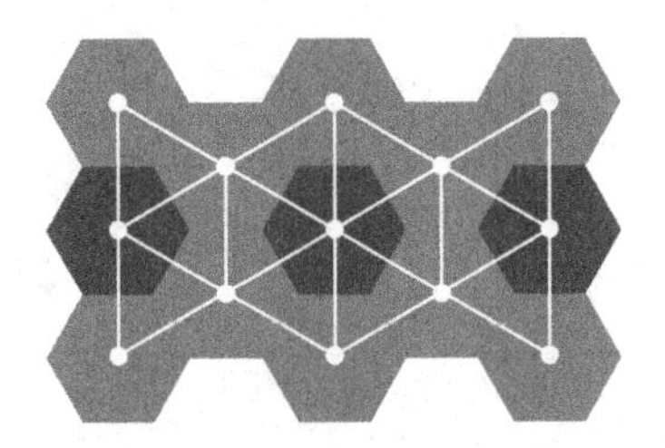

FIGURE 9.6 Cells of a GSM mobile phone network.

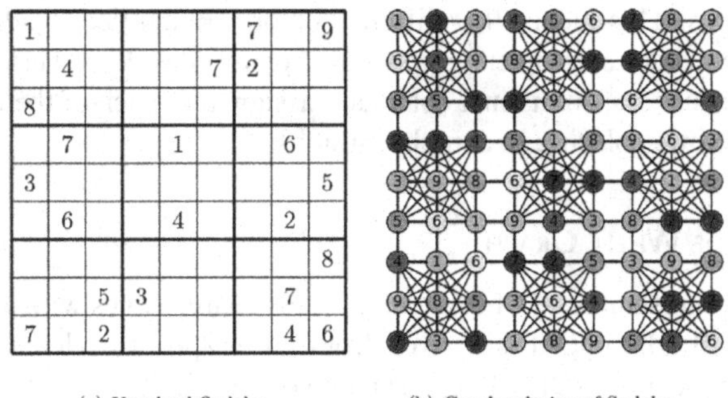

FIGURE 9.7 Unsolved Sudoku and solution obtained by graph coloring in Sudoku.

9.10.3 Sudoku Puzzle

The game of Sudoku is very popular among people. Graph coloring is used to solve Sudoku puzzles easily. The cells of the square grid are colored with different colors. For example, consider 2×2 square grid with 16 cells. Only four colors are needed to fill the 2×2 square cells.

Graph coloring is a specific instance of graph labeling in graph theory. It involves assigning labels—traditionally referred to as "colors"—to graph elements while adhering to certain restrictions. In its simplest form, vertex coloring refers to coloring a graph's vertices so that no two neighboring vertices share the same color. Similarly, in face coloring of a planar graph, a color is assigned to each face or region to ensure that no two faces that share a boundary have the same color. In edge coloring, a color is assigned to each edge so that no two adjacent edges share the same color.

Figure 9.7 evaluates graph coloring in Sudoku.

9.11 CALCULATING CNs FOR VARIOUS TYPES OF GRAPHS

This chapter covers the calculation of the CN for a variety of graphs, including the Cantor set, von Koch curve, Koch snowflake, Sierpinski arrow head curve, Hilbert curve, as well as complete, regular, star, cycle, and null graphs. The implementation of the related graphs and their structures determine the CN.

9.11.1 CN of Null Graph

A null graph consists of either one vertex or multiple vertices without edges. The CN of this graph corresponds to one of the specified types of graphs. Since no vertex is adjacent to another, each vertex is assigned a single color. Therefore, CN (G) = 1.

In the case of a null graph with only one vertex, it is colored with one color, leading to CN (G) = 1.

9.11.2 CN of Complete Graph

In a complete graph, every vertex has n edges, and there are n vertices in total. It is represented by the symbol Kn. Each vertex is assigned a distinct color. For a complete graph, CN(G) = n. This means that different colors are applied to adjacent vertices. For example, consider a complete graph with five vertices. Each vertex is connected to all the other vertices, so distinct colors are assigned to each of the five vertices.

Figure 9.8 shows coloring in complete graphs.

9.11.3 CN of Wheel Graph

The wheel graph resembles a wheel. Every vertex shares edges with a central vertex, making them all adjacent. It is derived from the graph of cycles, denoted as Wn. The wheel graph contains 2n − 1 edges and n vertices. In its first iteration, the wheel graph has four nodes and six edges. The first iteration features four nodes that are adjacent to one another, requiring four distinct colors. In the second iteration, the graph has five nodes and eight edges. The third iteration contains six vertices and ten edges. Each subsequent iteration adds one additional vertex and two more edges. The following formulas are used to determine the dominated CN.

$$CN(G) = \left\{ \begin{array}{l} 4 \text{ if no. of vertices is even} \\ 3 \text{ if no. of vertices is odd} \end{array} \right\} \tag{9.1}$$

The following table shows the CN in some iterations of the wheel graph (Table 9.2).

FIGURE 9.8 Chromatic number CN(G) obtained by coloring in complete graph.

TABLE 9.2
Chromatic Number of Wheel Graph

S. No.	Iteration	Graph	No. of Vertices	Chromatic Number CN(G)
1	1		4	4

(Continued)

TABLE 9.2 (*Continued*)
Chromatic Number of Wheel Graph

S. No.	Iteration	Graph	No. of Vertices	Chromatic Number CN(G)
2	2	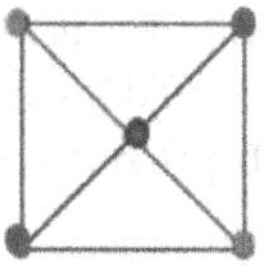	5	3
3	3	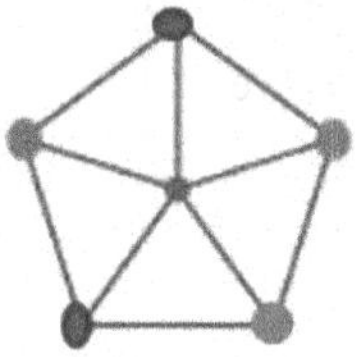	6	4
4	4	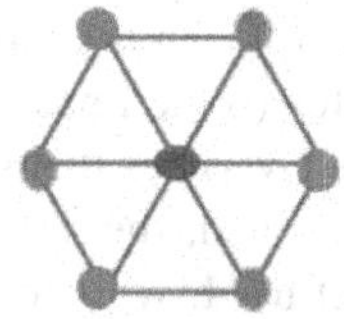	7	3
5	5	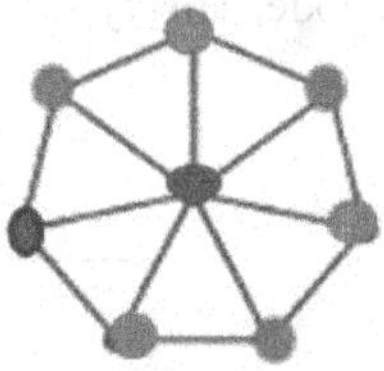	8	4
6	6	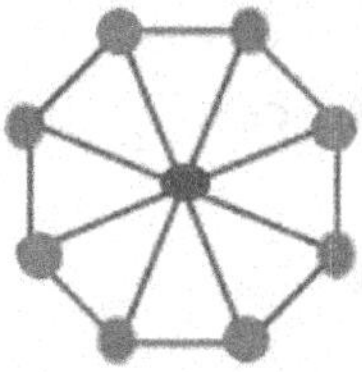	9	3
7	7	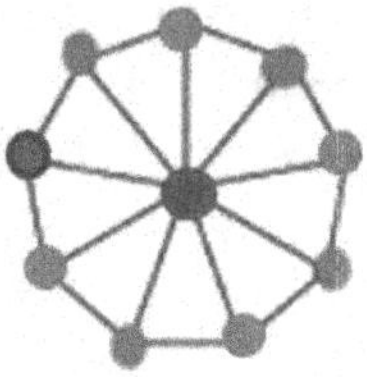	10	4
8	8	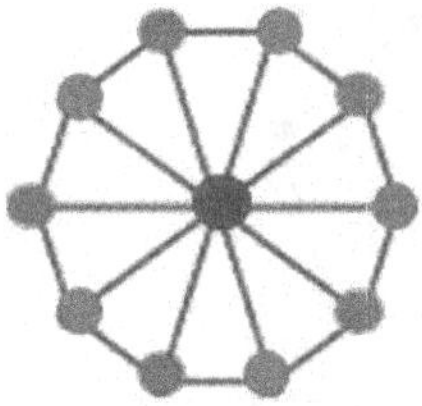	11	3

9.11.4 Chromatic Number of Cycle Graph

A cycle graph is a closed, linked graph where every cycle graph has the same number of edges and vertices. The path is closed, with non-intersecting edges connecting every node. Each vertex has a common degree of 2. The initial iteration of the cycle graph consists of three vertices and three edges. In the next iteration, both the number of vertices and edges increase by one. It is denoted by the symbol C_n. The CN can be determined using the following formula (Table 9.3)

$$CN(G) = \left\{ \begin{array}{c} 2 \text{ if n is odd} \\ 3 \text{ if n is even} \end{array} \right\} \tag{9.2}$$

9.11.5 CN of Star Graph

This type of graph piques my interest the most. In a star graph, the center vertex is connected to all other vertices. The notation for an n-degree vertex star graph is denoted as Sn, where the neighboring nth vertex (sometimes referred to as the center node) connects to s_n. $n-1$ vertices. These $n-1$ vertices themselves are not contiguous. Although the outer edges have been loosened, the structure still resembles a wheel graph. The center node has a degree of $n-1$, while every other vertex has a degree of 1 and can be considered a terminal node. This type of graph can be colored using just two colors, resulting in a CN of two.

TABLE 9.3
Chromatic Number of Cycle Graph

S. No.	Number of Vertices (n)	Cycle Graph Cn	$CN(G) = \left\{ \begin{array}{c} 2 \text{ if n is odd} \\ 3 \text{ if n is even} \end{array} \right\}$
1	3		3
2	4		2
3	5		3

9.11.6 CN of Equal Degree Graph

A graph with equal degree vertices is referred to as a regular graph or equal degree graph. By their degree on the graph, it is mentioned. For instance, a graph is referred to as m-regular if its order is m. Now let's begin coloring each vertex on the graph. The term "dominated chromatic number" refers to the smallest number of colors that are utilized to color the graph (G). The dominated CN of the Regular Graph is calculated by the formulae (Tables 9.4 and 9.5):

$$CN(G) = \left\{ \begin{array}{c} 3 \text{ if n is odd} \\ 2 \text{ if n is even} \end{array} \right\} \quad (9.3)$$

TABLE 9.4
Chromatic Number for 2 – Regular Graph

S. No	Degree of the Graph (m)	Number of Vertices V(G)	m-Regular Graph	CN(G)
1	2	3		3
2	2	4		2
3	2	5		3

TABLE 9.5
Chromatic Number for 3 – Regular Graph

S. No.	Degree of the Graph (m)	Number of Vertices V(G)	m-Regular Graph	CN(G)
1	3	7	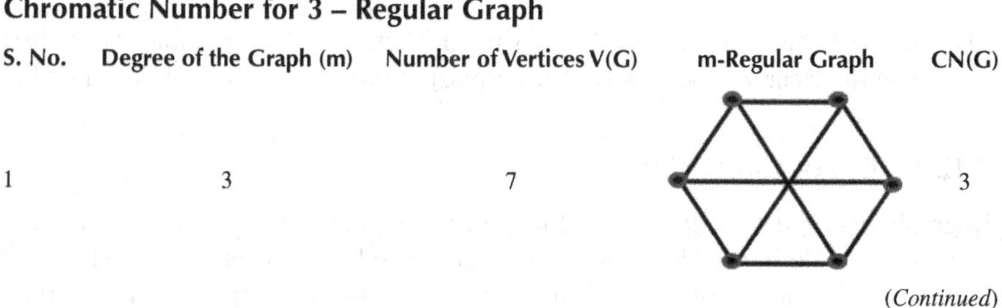	3

(Continued)

TABLE 9.5 (*Continued*)
Chromatic Number for 3 – Regular Graph

S. No.	Degree of the Graph (m)	Number of Vertices V(G)	m-Regular Graph	CN(G)
2	3	8		2
3	3	8		2
4	3	8		2

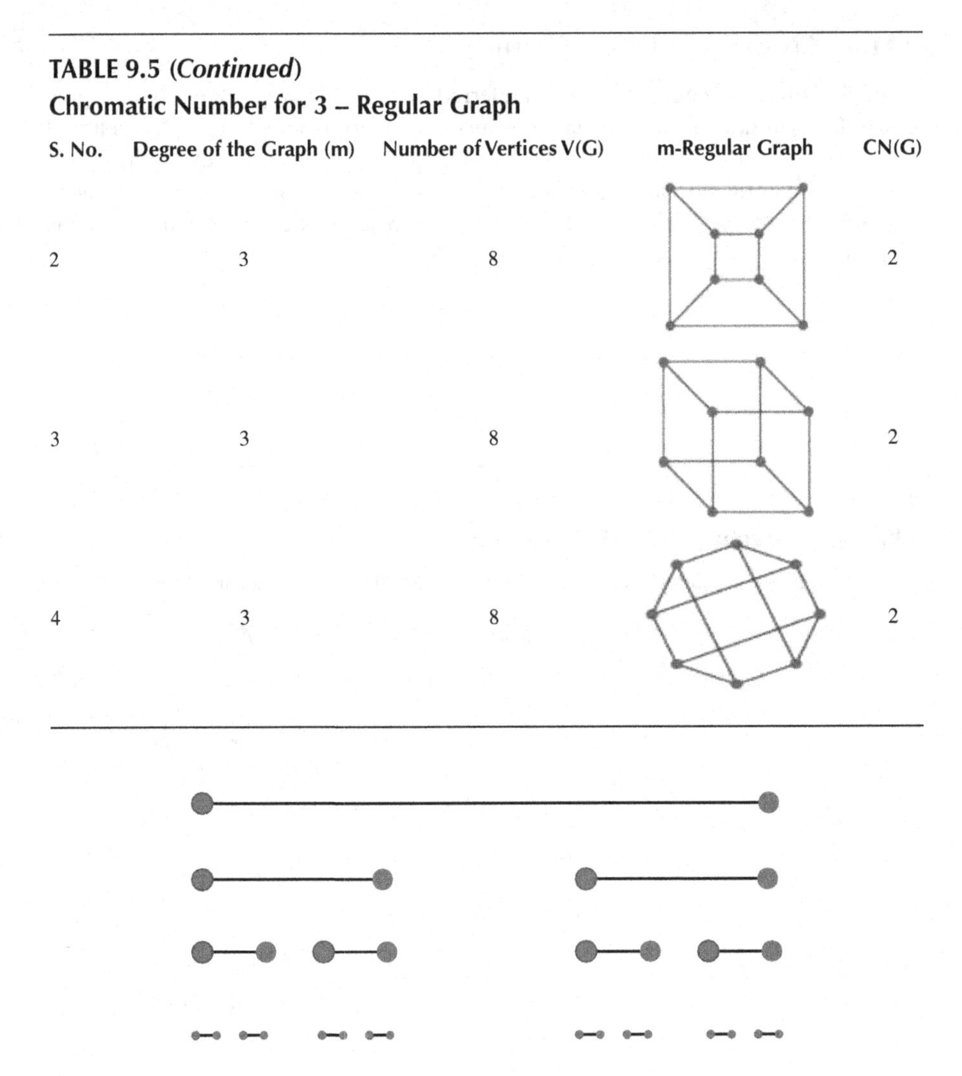

FIGURE 9.9 Graph coloring of Cantor set.

9.12 CN OF FRACTAL GRAPHS

This section focuses on the CN of self-similar fractal graphs, including the Hilbert curve, von Koch curve, Koch snowflake, Sierpinski arrowhead curve, and Cantor set [4].

9.12.1 CN of Cantor Set

In graph theory, the Cantor set is a fundamental example of fractal graphs. It plays a significant role in a wide range of applications today, contributing to fields such as computer science, animation, data structures, artificial intelligence, architecture, and more. The Cantor set begins with just one line, with an interval on units [0, 1].

By removing the middle third, a single line is split into three parts; the middle section is removed. In the original iteration, one edge splits into two. This process is repeated for every edge in the graph, resulting in each edge being further divided into two segments. Consequently, there will be twice as many edges in the next iteration.

It is merely colored with two hues; its CN is two. Every edge has a non-adjacent edge, and two distinct hues are present at their terminal vertices. Therefore, the von Koch curve is classified as a two-chromatic fractal graph.

9.12.2 CN of Von Koch Curve

Hedge von Koch made this discovery. In his research, he provided a geometric definition of fractal graphs. In 1904, he published a highly influential study on fractal graphs that motivated many mathematicians to focus on this area. It all began at the topological dimension, which is divided into three sections. The middle segment is straightened and bent outward to form the two sides of an equilateral triangle. For the next iteration, the same procedure is applied to every edge from the previous iteration. It takes the appearance of a path, with no intersections between the edges. One by one, the trail traverses each edge. Consequently, the von Koch curve can be colored using just two colors. This can be observed, as there are only two distinct colors at the alternative vertices.

Figure 9.10 shows the graph coloring of the von Koch curve.

9.12.3 CN of Koch Snowflake

The Koch snowflake is the next level of the von Koch curve. It is an early fractal graph and features a closed path with intersecting edges. Initially, equal-length triangles are used to start the first step, resulting in three lines at the beginning. The edges of the Koch snowflake at the mth level are given by the formula $(3 * 4)^m$. The Koch snowflake is the next level of the con Koch curve. It is an early fractal graph

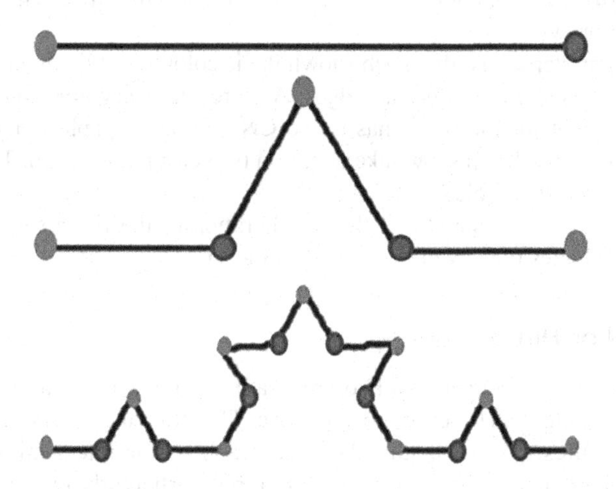

FIGURE 9.10 Graph coloring of von Koch curve.

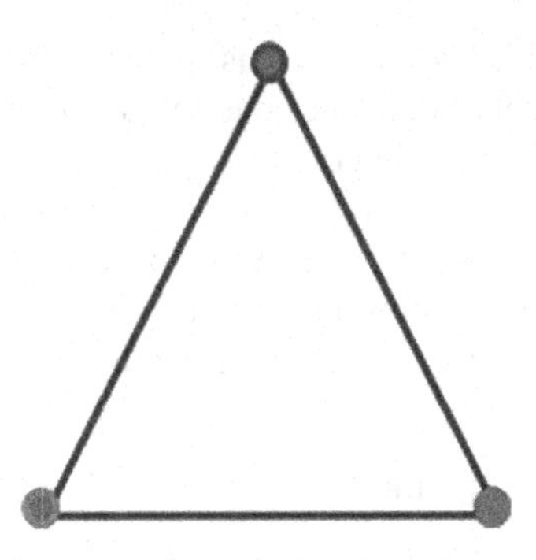

FIGURE 9.11 Chromatic number of first iteration of Koch snowflake.

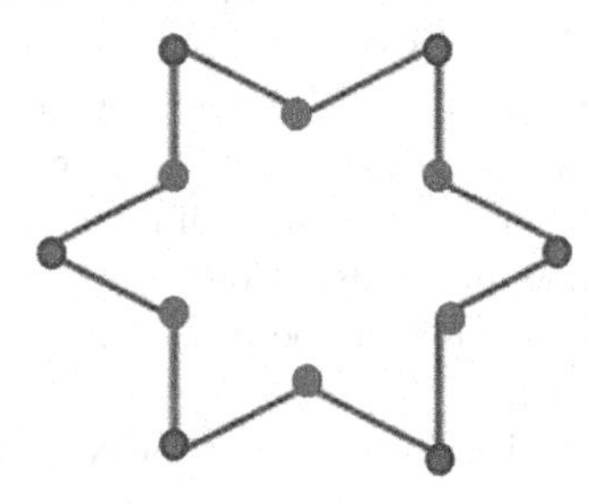

FIGURE 9.12 Chromatic number of second iteration of Koch snowflake.

and features a closed path with intersecting edges. Initially, equal-length triangles are used to start the first step, resulting in three lines at the beginning. The edges of the Koch snowflake at the mth level are given by the formula $(3*4)^m$, where m is a positive integer that begins at zero and increases by one. Except for the first iteration, all lines are even. The first iteration contains three edges and three nodes, and it features different colors at each vertex. For example, the vertices may be colored red, blue, and brown.

In subsequent iterations, the Koch snowflake is colored in two distinct hues, with different colors applied to alternate edges. As a result, every iteration of the Koch snowflake—except for the first—has a two-CN colorable graph. For instance, the initial iteration of the Koch snowflake depicts a two-chromatic graph. The other vertices are colored red and blue.

Figure 9.11 evaluates the CN of the first iteration of the Koch snowflake, while Figure 9.12 evaluates the CN of the second iteration.

9.12.4 CN of Hilbert Curve

This is an additional instance of a two-chromatic fractal graph, with the most valuable structure being the Hilbert curve structure. This fractal graph exhibits self-similarity and features an open path. The edge graph is non-intersecting. Numerous real-world applications exist for this fractal graph, particularly in fractal antennas. The initial iteration consists of four vertices and three edges, with alternative vertices

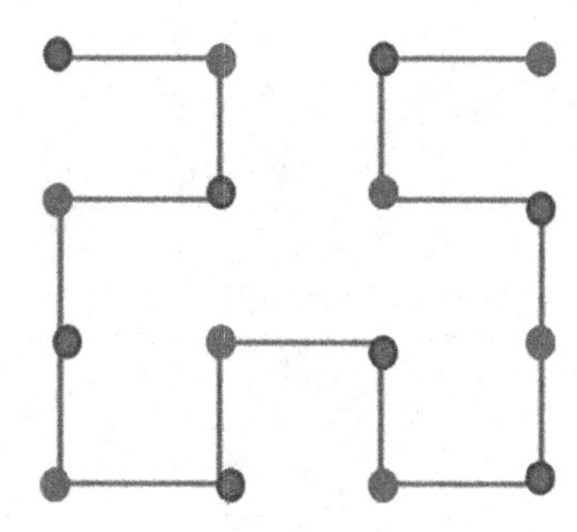

FIGURE 9.13 Coloring of second iteration of Hilbert curve.

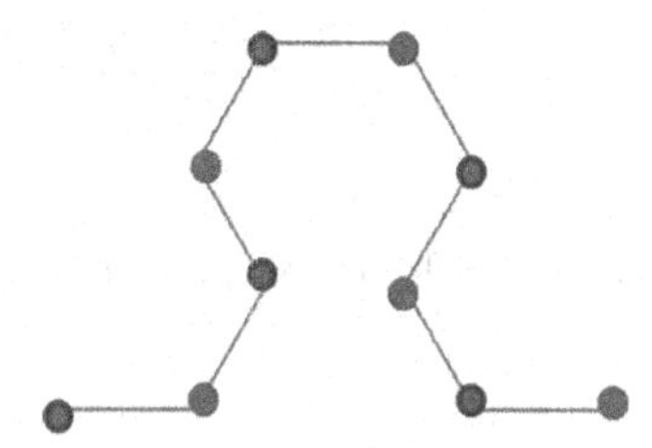

FIGURE 9.14 Coloring of second iteration of Sierpinski arrowhead curve.

colored uniquely. In the next iteration, the number of nodes and edges will increase in a constant ratio. The Hilbert curve has 4^n vertices for the nth iteration.

Figure 9.13 illustrates the Hilbert curve's second iteration, which is also a two-chromatic graph.

9.12.5 CN of Sierpinski Arrowhead Curve

The Sierpinski arrowhead curve is another example of an open path fractal graph, similar to the Hilbert and von Koch curves. It is a continuous fractal graph sequence defined recursively, characterized by its curvature that fills space. It begins in the form of a hexagon. In the next iteration, each edge is converted to a 60° angle, with the direction shifting from left to right, while the terminal vertices remain stationary. For every Sierpinski arrowhead curve iteration, the number of vertices is even. It is an open path graph that does not overlap. Typically, the Sierpinski arrowhead curve has $3n+1$ vertices and 3n edges in its nth iteration. Alternative vertices are assigned different colors, such as blue and red.

Figure 9.14 shows the Sierpinski arrow's second iteration.

9.13 APPLICATION OF COLORING WITH FRACTAL GRAPH

Coloring is an integral part of our lives. It is something we use every day without even realizing it. This section explores how coloring is intertwined with our existence in various ways [5].

FIGURE 9.15 Fractals in nature: An identical shape to make it beautiful.

9.13.1 Nature

God created this world, designing beautiful images in every object. He gave identical shapes to enhance their beauty and stitched uniform patterns into them. In reality, fractal graphs can be observed everywhere—in clouds, mountains, rivers, and more. He crafted the atoms and cells of every living thing and their functions in the same manner. Every cell in the human body possesses unique characteristics. Beautiful images become even more striking with color, which follows a similar pattern across different forms. Some examples include leaves that are generally green, mountains that may also appear green, human blood cells that are red, and clouds that are white. Each tree exhibits variations within its species. Around the world, nature shows this consistent pattern with minimal differences, depending on the circumstances.

Figure 9.15 illustrates an example of fractals in nature.

9.13.2 Atom

An atom is a very small unit into which matter can be divided without the liberation of electrical charges. It possesses the characteristics and properties of a chemical element. Most of an atom is empty space. The remainder consists of protons, which are positively charged, and neutrons, which are electrically neutral. Each part of an atom is distinguished by different colors. The coloring of an atom helps to identify its parts individually, making it easier for readers who are exploring atomic structure for the first time to differentiate its components. Figure 9.16 illustrates the structure of an atom.

9.13.3 Man-Made Objects

Fractals are the study of objects that exhibit self-similar patterns. Humans create beautiful self-similar patterns in their surroundings, such as kolam designs, wall

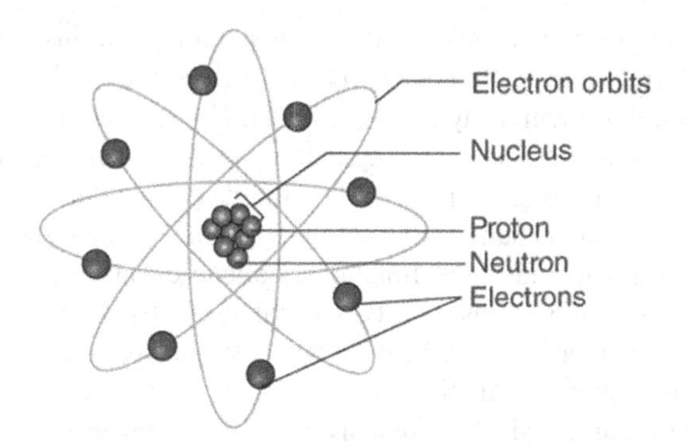

FIGURE 9.16 Atom.

FIGURE 9.17 Floor wallpaper with fractal design.

decorations, window structures, door patterns, floor wallpapers, and more. People take delight in these beautiful self-similar forms throughout their lives, as shown in Figure 9.17.

9.13.4 Rangoli

Rangoli is a cultural art form discovered in Indian culture, especially within Tamil culture. It is created using materials such as fine powdered limestone, powdered rice, boiled or unboiled rice flour, colored sand, quartz powder, and rocks of various colors. It serves as a holy identification for each Hindu household in everyday practice. Rangoli represents joyfulness, positive vibrations, and the holiness of the home. The main theme of Rangoli revolves around decoration. Traditionally, people have used materials like calcite powder, limestone powder, and red stone powder to create

Kolam. Limestone has the capacity to kill germs and prevent insects from entering the household. Cereal powders made from rice or wheat are used to provide food for small insects, keeping them away from the home. Hindus believe Rangoli serves to welcome Lord, who is the deity of living happily ever after in a wealthy and auspicious manner. Rangoli is eagerly preferred by Indian people during celebrations like Deepavali, Onam, Pongal, Sankranti, and many other Hindu festivals, particularly in Tamil Nadu. During traditional functions, college events, and various religious festivals, Rangoli is displayed as an artwork that symbolizes beauty and happiness. Nowadays, many competitions are held for Rangoli, showcasing creativity, colorfulness, and brightness at most functions.

Rangoli designs can consist of simple as well as complex geometric shapes, illustrations of goddesses and immortals, flowers, and shapes related to the specific celebrations. It is created using numerous elaborate designs and features self-similar geometric patterns in all layers of Rangoli, resembling a fractal graph. The coloring of Rangoli is also highly enthusiastic. Rangoli is colored in adjacent regions using various colors. Non-adjacent regions are colored the same, while adjacent areas are colored differently.

Figure 9.18 shows how Rangoli is colored in adjacent regions using various colors. Non-adjacent regions are colored the same, while adjacent areas are colored differently.

9.14 COLORING IN ARCHITECTURE

Graph Theory has introduced many advanced tools for research areas such as Matching, Domination, and Coloring. It has many applications in various fields, including Architectural Engineering and Artificial Engineering. Vertex Coloring refers to the process of assigning colors to each vertex according to specific rules. The rules state that all vertices must be colored, and adjacent vertices cannot share the same color. Edge terminals are colored with different colors. This approach has numerous computer-related applications. This paper explains the application of Graph Coloring in real-life fractal structures.

Graph coloring has many applications in real life and architectural areas. These include guarding art galleries, layout segmentation, scheduling of round-robin games, aircraft seedling, bi-processor tasks, frequency assignment problems, timetable

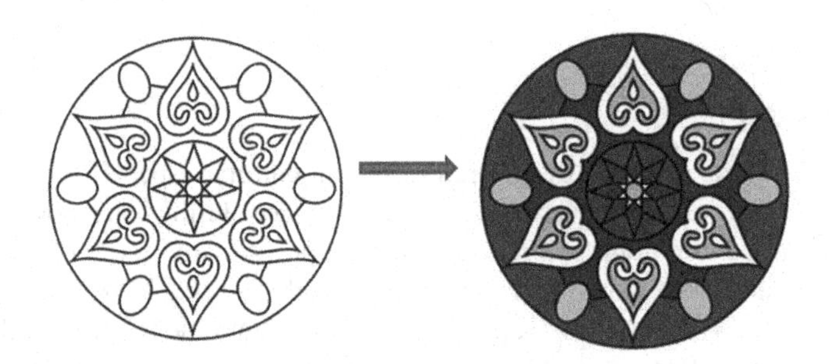

FIGURE 9.18 Fill color with rangoli design.

FIGURE 9.19 Ceiling of hall of 1000 pillars.

problems, map coloring, and GSM phone networks, among others. It simplifies complex puzzles and critical problems. It is evident that graph coloring is an advanced tool for research in the upcoming years. For instance, two rows of pillars with yali motifs carved on them can be found at the historic Meenakshi Nayakkar Mandapam, also known as the "Hall of 1000 Pillars." Built in 1569 by Ariyanatha Mudaliar, the hall combines technical prowess with creative vision.

The decorative pattern of the temple resembles Rangoli fractal structures with graph coloring [6]. The given It is an example of fractal structures with graph coloring, which can be seen on the ceiling of the Hall of 1000 Pillars in the Madurai Meenakshi Amman Temple located in Madurai city, India.

Figure 9.19 is an example of fractal structures with graph coloring, which can be seen on the ceiling of the Hall of 1000 Pillars in the Madurai Meenakshi Amman Temple located in Madurai city, India.

9.15 FRACTALS

The excitement people feel for beautiful self-similar images is what defines fractals. Long before the invention of computers, people were working on fractals. They originated from efforts to measure the length and boundary of the British coast. They used large-scale maps that represented only half the length of the coastline measured on a proper map. Consequently, mathematicians introduced fractals as powerful tools for measuring complex phenomena, solving problems in computer science, and uncovering secrets from a wide variety of systems.

9.16 APPLICATION OF COLORING WITH FRACTAL RANGOLI GRAPH

Coloring is an integral part of our lives. We engage in coloring every day, often without realizing it. This section explores how coloring is interwoven with our lives in various ways.

9.16.1 Creature of Rangoli or Kolam

Kolam is an ancient traditional decorative art that adds a sense of culture to our homes, typically placed at the entrance. It is the origin of Graph Theory as well as Fractal Graphs [7]. This art form originates from ancient Tamil Nadu, which is one of the states in India, but it has since spread to other southern Indian states and around the world. Kolam consists of geometric line drawings made from straight lines, curves, loops, and grid patterns of dots. The coloring of these geometric line drawings follows specific guidelines throughout the Rangoli or Kolam [8]. The structure of these drawings embodies the self-similarity property. Various regional versions of Kolam, each with their own distinctive forms, reflect the diversity of our culture.

9.16.2 Using Material of Rangoli

Rangoli is a cultural art form that emerged from Indian culture, particularly Tamil culture. It is created using materials such as fine powdered limestone, rice flour (both boiled and unboiled), colored sand, quartz powder, and various colored stones.

9.16.3 Purpose of Rangoli

Rangoli serves as a sacred symbol for each Hindu household in their daily practices. It represents joy, positive vibrations, and holiness within the home. The main theme of Rangoli revolves around decoration. Traditionally, materials such as calcite powder, limestone powder, and red stone powders have been used to create Kolam. Limestone possesses germicidal properties, helping to keep insects at bay. Cereal powders made from rice or wheat are placed to provide food for small insects, preventing them from entering the household. Hindus believe that Rangoli is intended to welcome deities who bring happiness, prosperity, and auspiciousness. Furthermore, Rangoli is particularly favored by Indian people during celebrations like Deepavali, Onam, Pongal, Sankranti, and many other Hindu festivals, especially in Tamil Nadu, as well as during traditional functions, college events, and various religious festivals.

9.16.4 CN of Rangoli Fractal Graph

Rangoli designs are created using both simple and complex geometric shapes, illustrations of goddesses or immortals, flowers, and husk shapes that relate to the specific celebrations. They often feature an elaborate array of designs and are characterized by self-similarity in their geometric patterns, resembling a fractal graph. The coloring of Rangoli is quite enthusiastic. Rangoli, demonstrating how to color adjacent regions with various colors. Non-adjacent regions are colored the same, while adjacent areas are colored differently.

The following Rangoli can be colored using graph coloring techniques. Figure 9.18 exhibits a self-similar property, which may be horizontal, vertical, or diagonal. It is composed of many layers. The outer layer is colored brown and occupies a large surface area in the Rangoli, so it does not repeat in the other parts. The inner layer is hexagonal and should be colored blue. An oval shape is positioned at each corner of the hexagon, equidistant from each other. This oval shape lies between the circle of the outer layer and the hexagonal inner layer, so it should be colored in an alternating color, excluding brown and blue. Thus, it is colored yellow. The center of the hexagon is filled with a yellow heart, which conveys love; yellow is often used to express liking and friendship, as well as joy and happiness. The inner third layer is circular and filled with lavender. It also exhibits self-similar drawings. The drawing structure consists of rectangle-based duplex triangles, meaning the width of the rectangle is joined with identical-length equilateral triangles. There are three different identical structures, each colored differently: pink, white, and yellow.

9.17 CONCLUSION

Kolam is a form of traditional decorative art. Rangoli, which is part of this art form, showcases beauty and happiness. Nowadays, many competitions are conducted for Rangoli, which is characterized by its informative nature, vibrancy, and brightness during various functions. These beautiful fractal Rangoli structures have been incorporated into architecture. Today, architectural engineers prefer to build structures that include numerous fractal designs, enhancing their aesthetic appeal through coloring based on regional graph coloring rules. Future work will analyze fractal pattern design with graph coloring in some of the world's most ancient temples, such as the Ellora Cave Temple and Konark Sun Temple. Coloring, a sophisticated technique in graph theory, has garnered significant interest from mathematicians, leading to numerous practical applications in everyday life. The CN applies to self-similar fractal graphs and various graphs in this chapter. Essentially, a CN signifies the minimum number that can be used for non-adjacent graphs. The two chromatic graphs along that path are described in this chapter. Applications span a wide array of fields, including computers, data structures, the traveling salesman problem, data encoding, image compression, screenplay background setting, image display, resource allocation, process planning, and more. It represents the simplest method of labeling a graph, which includes nodes, edges, and potentially even a section or division of a planar graph.

REFERENCES

1. Ore, O., *The Four Color Problem*, Academic Press, New York (1967).
2. Guptha, P., and Sikhwa, O., A study of vertex-edge coloring techniques with application, *International Journal of Core Engineering & Management*, Vol. 1(2), pp. 27–32 (2014).
3. Balakrishnan, R., Wilson, R. J., and Sethuraman, G., *Graph Theory and Its Applications*, Narosa Publications (2015).
4. Falconer, K. J., *The Geometry of Fractal Sets*, Cambridge University Press, Cambridge, 1985.
5. Guichard, D. R., No-hole k tuple, (r+1) distant colorings, *Discrete Appied Mathematics*, Vol. 64(1), pp. 87–92 (1996).
6. Gyarfas, A., and Lehel, J., Online and first fit coloring of graphs, *Graph Theory*, Vol. 12(2), pp. 217–227 (1988).
7. Harary, F., *Graph Theory*, Addison Wesley Publishing Company, Reading, MA (1969).
8. Mahapatra, T., Ghorai, G., and Pal, M., Fuzzy fractional coloring of fuzzy graph with its application, *Journal of Ambient Intelligence and Humanized Computing*, Vol. 11(11), pp. 5771–5784 (2020).

10 Fractals in Medicine

10.1 INTRODUCTION

Fractals play a crucial role in enhancing medical research, diagnosis, treatment, and the provision of healthcare across various fields and specializations. Their unique mathematical properties pave the way for innovative methods of treating patients and managing diseases by providing valuable insights into the structure and complexity of biological systems. Fractals are applied in areas such as medical imaging, physiological processes, biological structures, and clinical decision-making. The intricate structures found in biological systems, including the branching patterns of blood vessels, bronchial tubes, and neural networks, are analyzed using fractals [1]. By quantifying the fractal features of these structures, researchers can gain a deeper understanding of their formation, organization, and function. For instance, studying the fractal patterns of blood vessel networks can illuminate processes such as angiogenesis, vascular remodeling, and diseases like diabetic retinopathy and cancer [2].

Medical imaging techniques, such as MRI, CT scans, and ultrasounds, utilize fractal-based algorithms to enhance image quality, segmentation, and feature extraction. Fractal analysis in medical imaging can assist in the diagnosis and monitoring of various conditions, including cancers, neurological disorders, and cardiovascular issues [3]. Additionally, it aids in characterizing tissue properties and detecting anomalies by analyzing texture patterns in medical images. Fractals are also employed to investigate and simulate physiological data, including electroencephalography, breathing patterns, and heart rate variability. Fractal analysis can uncover the underlying dynamics, autonomic control systems, and pathological changes associated with conditions such as heart failure, epilepsy, and sleep disorders. In clinical practice, fractal-based biomarkers derived from physiological signals are valuable for both prognostication and diagnosis. The creation of nanoparticles, microstructures, and implantable drug delivery devices is studied using fractal geometry [4]. By enhancing tissue targeting, release kinetics, and medication distribution, fractal analysis contributes to the safety and effectiveness of therapeutic interventions. To guide medication development and optimize dosing regimens, fractal-based models are also utilized to predict drug absorption, distribution, metabolism, and excretion. Fractals serve as useful tools for modeling and simulating intricate biological processes, such as disease development, infection transmission, and responses to therapies. These models help explore disease mechanisms, identify therapeutic targets, and evaluate treatment options by incorporating the multiscale nature of physiological processes. Fractal simulations predict specific patient outcomes and optimize clinical interventions, thus supporting personalized medicine.

DOI: 10.1201/9781003481096-10

10.1.1 Tumor

Tumors generally develop when the body's cell division process is disrupted. Normally, cell division is tightly regulated, with older cells being replaced by new ones, or new cells being created to take on different roles. Tumors can arise when the equilibrium between cell division and cell death is upset, which can occur due to issues with the immune system. Tumors are classified as either benign or malignant, with growth being a crucial factor in tumor development. We assess this growth by examining the cell population.

Figure 10.1 shows the tumor in other parts of the organ, while Figure 10.2 shows the different types of tumors.

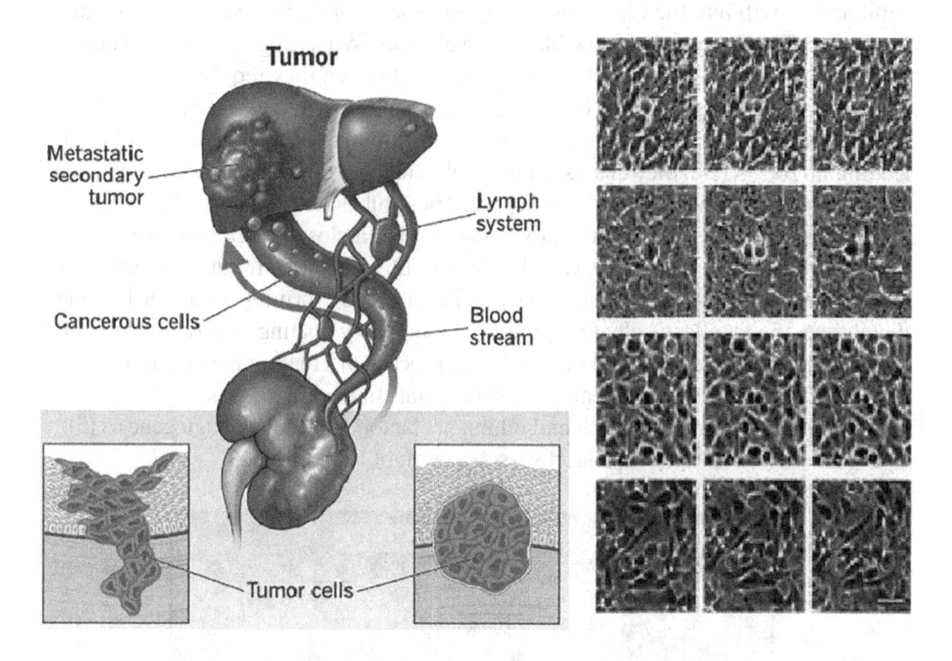

FIGURE 10.1 Tumor on organ.

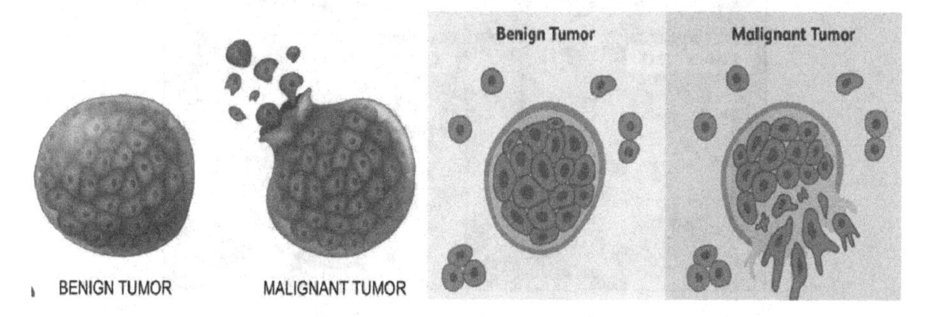

FIGURE 10.2 Types of tumor.

10.2 MODELING OF TUMOR GROWTH USING FRACTALS

One method to illustrate the intricate structure of biological systems is through mathematical modeling of tumor growth. The combination of fractal dimension and percolation generation produces a model that depicts tumor growth. Fractal geometry has been applied to the biological sciences, significantly enhancing our understanding of the complex morphological characteristics and functional aspects of cells and tissues.

Angiogenesis, the physiological process of creating new blood vessels from pre-existing ones, is essential for both wound healing and growth. However, it plays a crucial role in the transformation of tumors from a dormant to a malignant state. Tumors induce blood vessel formation by secreting various growth factors, such as basic fibroblast growth factor and vascular endothelial growth factor, which promote capillary growth into the tumor, providing necessary nutrients and facilitating tumor expansion. The formation of new blood vessels is known as angiogenesis, and tissues release chemicals to stimulate blood vessel growth when they require more oxygen.

The process of angiogenesis begins with the breakdown of the basement membrane, followed by the departure of endothelial cells from the vessel. As the cells behind the leading tip proliferate, they coalesce into a tube that "sprouts" from the original capillary, extending further into the wound space. These tubes may branch at their terminals and eventually unite with additional sprouts to form a closed loop, allowing blood to flow through. From these new vessels, the sprouting process reinitiates until a network of new capillaries permeates the wound area. Before forming a genuine basement membrane, the capillary cells first produce a temporary coating of proteoglycans and fibronectin. Angiogenesis, the body's natural process of creating new capillary blood vessels for reproduction and healing, is significant. In healthy tissues, the body maintains a perfect balance of growth and inhibitory factors to regulate angiogenesis [5].

Figure 10.3 illustrates the blood vessel growth of the tumor.

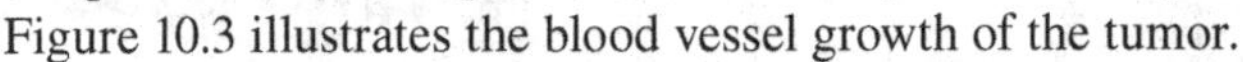

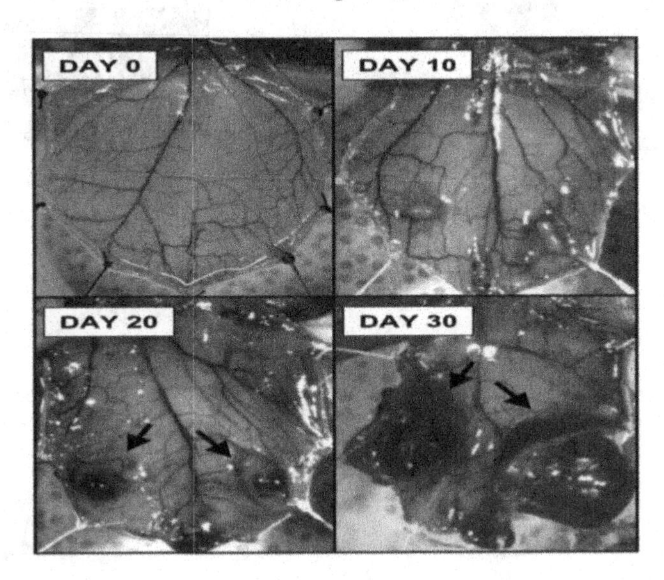

FIGURE 10.3 Blood vessel growth of tumor.

Self-similarity becomes an even more ill-defined concept in the real world. As a result, structures are typically described as being "fractal-like" or modeled using truncated fractals [6]. While real structures and segments are comparable, real structures have a "thickness" that introduces some "noise" to the ideal fractal structure they resemble. Additionally, there are minimum and maximum values that limit the scaling behavior, which must satisfy self-similarity. The structure itself has the potential to further diminish this concept. This is a crucial topic for biological systems, which Brown and colleagues have recently studied using scaling connections. By introducing a scale according to which the original length of segmentation is partitioned, and counting the number N of self-similar parts resulting from the partitioning, the fractal dimension D is defined as follows:

$$D = \log_e N. \tag{10.1}$$

Two intrinsic limits—the picture's resolution and dimension—determine the minimum (ε_m) and maximum (ε_M) values for the scale parameter when considering the image of a self-similar structure. A workable method for estimating D when considering the image of fractal structure would be to choose a range of values $\varepsilon(\varepsilon_m < \varepsilon < \varepsilon_M)$, for each tessellation of the image in boxes and figure out N_b how many boxes contain the image structure [Box Counting Method], which was first presented by Mandelbrot. The evaluation of the linear regression and the tessellation process are obviously related to the results. Previous research on self-similar architectures found in many biological structures, including the bronchial tree and the placental villous tree, as well as the occurrence of scaling relationships—that is, when two variables, X and Y, are related according to a specific power law—provide justification for an approach based on the assessment of fractal properties [7].

$$Y = Y_0 X^p \tag{10.2}$$

where X is the tumor mass, Y is the basal metabolic rate, Y_0 is the normalization constant, and p is a non-integer variable that varies depending on the stage at which the tumor is forming. Here, Fractal Dimension is used to model the tumor's growth. Several techniques are employed to determine the tumor's dimensions, and growth models for the tumor are described [8]. Tumors generally seem to result from aberrant control over cell division. The body's ability to divide cells to produce new ones or to carry out new tasks is normally tightly regulated. Damaged or unnecessary cells pass away to make room for healthy ones. The cell growth mathematical modeling is provided in the next section.

10.3 MATHEMATICAL MODELING OF CELL

Millions of cells make up the human body. Normally, there is an orderly division and multiplication of these cells. As new cells replace old ones, our growth pattern is shaped by our genes and lifestyle choices. Every living organism undergoes a continuous process of cell division and regeneration. With the exception of viruses, the cell is the fundamental unit of structure and function in all living things.

The quantitative study of cell proliferation has significantly enhanced our understanding of cell functions. Changes in the number of unicellular microorganisms within a colony typically indicate its growth, as opposed to changes in the size of the individual organisms. The process of binary fission, where a cell divides into two, is a common method for cell multiplication and proliferation. Numerous methods exist for determining the number of cells in a culture, including counting the number of cells per unit volume, measuring the mass of the cells, and analyzing the biochemical activity of the cells [9].

According to the law of mass action, if the population N were doubled, the change ΔN would also be doubled. So it can be written as

$$\Delta N = KN\Delta t \tag{10.3}$$

where K is proportionality constant.

If we divide the Equation 10.4 on both sides by Δt and take the limit as $\Delta t \rightarrow 0$, we obtain the derivative. Thus

$$\underset{\Delta t \rightarrow 0}{\text{Lt}} \frac{\Delta N}{\Delta t} = KN \tag{10.4}$$

$$\text{i.e.,} \frac{dN}{dt} = KN \tag{10.5}$$

It is expected that this relationship, which exists between the amount N and its derivative, will always exist. A differential equation is what we term such an equation. Using our knowledge of well-known functions and their derivatives to our advantage is the most basic approach to solving problems. Because we are familiar with the exponential function, we might try to solve the problem.

$$N = ce^{Kt} \tag{10.6}$$

where c is a constant. The meaning of the constant becomes clear if we specify the initial value N_0 of our population at time $t = 0$,

$$N_0 = ce^{K0} = c, \tag{10.7}$$

so that the constant "c" represents the initial value of the population. The solution of this equation may therefore be written as

$$N = N_0 e^{Kt} \tag{10.8}$$

It is demonstrable that no other function fulfils both the beginning condition and the differential equation $N(0) = N_0$. It follows that the solution is deemed unique. We can always get the population number N from the solution. For an equation like Equation 10.5 to have any significance at all, it must first be dimensionally correct. As we can see from the previous explanation, if we let the bracket [] to represent "the dimensions of" the quantity contained within them,

$$\left[\frac{dN}{dt}\right] = \left[\frac{\Delta N}{\Delta t}\right] = \left[\frac{N}{t}\right] \tag{10.9}$$

From Equation 10.5, the dimensions of dN/dt must be the same as $[KN]$, and therefore $[K] = [1/t]$, or, K has the dimension of reciprocal time. We can also recognize this fact from Equation 10.8, because the argument of the exponential function or exponent of e must be a pure number, whence $[Kt] = [1]$. From Equation 10.5 we see that the fractional growth rate at any time $\left(N^{-1}dN/dt\right)$ is a constant, and this constant is K. It is also called the specific growth rate.

Doubled exponential growth is a type of cell population expansion, meaning that the number of cells in each generation should be double that of the previous generation. However, not every cell survives through each generation, so the total number of generations only provides a maximum estimate. Under normal circumstances, the cells of the human body proliferate and expand in an ordered and regulated manner. One way to express cell development is through "transformation," an operation that changes a mathematical expression into a different but still valid form. Laplace transformation can be used to represent the transformation of cell growth since it satisfies two requirements: It must be in exponential order and it should be continuous or piecewise:

$$\text{i.e.,} L[f(t)] = \int_0^\infty e^{-st} f(t)\,dt, \quad t > 0 \tag{10.10}$$

Radial distribution functions (RDFs), denoted as g(r) can be used to analyze cell proliferation. An RDF is a mathematical tool used to characterize a group or collection of objects. It is conceptually defined as the observed probability of finding an object in a specific region divided by the probability of finding an object in that region if all the objects were uniformly distributed. Therefore, g(r) serves as a local indicator of how closely the observed distribution resembles a uniform distribution. g(r) is mathematically defined as

$$g(r) = \frac{dN/N}{dV/V} \tag{10.11}$$

$$= \frac{dN}{dV} \cdot \frac{V}{N} \tag{10.12}$$

where dN and dV stand for the number of objects in and volume of the sub-region (a tiny local area inside the system) under study, respectively, and N and V for the number of things in and volume of the overall system, respectively. The location vector from the system's center to the subregion of interest is represented by the vector r. The particles in the sub-region have a uniform distribution when g(r) equals one. Values that are higher or lower than one indicate an increased or decreased probability in comparison to a uniform distribution. It explains how the distance from a specific spot affects the density of the surrounding stuff. More precisely if there is cell at the origin O, and if $n = N/V$ is the average number density, then local density at distance r from O is given by $ng(r)$.

Suppose, for example, we choose a cell at some point O in the volume. If $\rho = N/V$ is the average density, then the mean density at P given that there is a cell at O would differ from ρ by some factor $g(r)$. So one could say that the radial distribution function takes into account the correlations in the distribution of cells arising from the forces that the cells exert on each other. Mean local density at distance r from $O = \rho g(r)$.

As long as the tumor in any part of the organ, the correlations in the position of the cells that $g(r)$ takes into account are due to the potential $\phi(r)$ that a cell at P feels owing to the presence of the cell at O. Using the Boltzmann distribution law

$$g(r) = e^{-\phi(r)/kt} \tag{10.13}$$

If $\phi(r)$ was zero for all r, i.e., if the cells did not exert any influence on each other, then $g(r) = 1$ for all r. Then mean local density would be equal to the mean density ρ: the presence of a cell at O would not influence the presence or absence of any other cell and the tumor would be normal. As long as there is an $\phi(r)$ the mean local density will always be different from the mean density ρ due to the interactions between the cells. Humans are comprised of trillions of cells that are organized into tissues such as muscle and skin and organs like the liver or lung. The proper function of human bodies is dependent on smaller structures, or organs, such as the heart or lungs. The tiny cells making up these organs actually contain within them Trillions of cells make up the body of a human, arranged into organs like the liver and lung and tissues like muscle and skin. Smaller organs like the heart and lungs are essential to the human body's correct operation. Smaller structures called organelles are actually contained within the microscopic cells that make up these organs. The functions of the cells are aided by these organelles. Many of the cells that comprise the body age and die throughout the course of a lifetime. The body has to replenish these cells in order to continuc operating at its best. Organelles are tinier structures. The functions of the cells are aided by these organelles. Many of the cells that comprise the body age and die throughout the course of a lifetime. The body has to replenish these cells in order to continue operating at its best.

10.4 MATHEMATICAL MODEL OF INFECTIOUS DISEASE

When an infectious disease (or tumor growth) can persist in a population without the need for outside assistance, it is referred to as endemic. This indicates that each infected cell, on average, spreads the infection to exactly one additional cell; any more, and the number of diseased individuals would increase exponentially, resulting in an epidemic; any fewer, and the illness will eventually disappear.

In mathematical terms, that is:

$$R_0 \times S = 1 \tag{10.14}$$

Assuming that everyone is susceptible, the basic reproduction number (R_0) of the disease multiplied by the percentage of the population that is truly susceptible (S) must equal one (because those who are not susceptible do not appear in our calculations because they are unable to contract the disease). Observe that this relationship implies that, in order for a disease to exist in an endemic stable state, the proportion of the population that must be susceptible must be smaller the higher the basic reproduction number, and vice versa. Let us assume, for the purposes of this discussion, that the population as a whole lives to age L and then passes away. If A is the average age of infection, then, generally speaking, people under A are susceptible, and people beyond A are immune.

$$S = \frac{A}{L} \tag{10.15}$$

$$S = \frac{1}{R_0} \tag{10.16}$$

$$\frac{1}{R_0} = \frac{A}{L}$$

$$R_0 = \frac{L}{A} \tag{10.17}$$

This gives us a straightforward method to calculate the parameter R_0 from readily accessible data. It seems that for a population whose age distribution is exponential,

$$R_0 = 1 + \frac{L}{A} \qquad (10.18),$$

where, the fundamental reproduction number is R_0. The average number of people that an infected person will typically infect in a community lacking immunity to the illness.

S: The proportion of the population that is susceptible to the disease because they are either immune or diseased.
A: The average age at which a population becomes ill.
L: The average longevity of a particular group of people.

Therefore, mathematics is required to calculate this slightly more complicated. That being said, given A and L, this does allow us to find the fundamental reproduction number of a disease for any type of population distribution. This enables us to identify if the tumor is cancerous or benign [10].

10.5 COMPUTATION OF FRACTAL DIMENSION

The tumor image is seen as two-dimensional, with coordinates denoted as (x, y). After dividing the (x, y) coordinates into grids that measure, the picture pixel size in the tissue is represented by s. If the minimum and maximum binary image levels lie inside the and boxes, respectively, then will contribute to the (i, j)th grid. Grid is defined as $n_r(i, j) = 1 - (k - 1)$

$$n_r(i,j) = 1 - k + 1 \tag{10.19}$$

In this method $N(r)$ is defined in this manner as the total of the contributions from each grid that is present in an image window.

$$N(r) = \sum_{i,j} n_r(i,j) \tag{10.20}$$

fractal dimension can be estimated as the slope of the line that best fits points $(\log(1/r), \log N(r))$.

$$\log(N(r)) = \log K + D\log(1/r) \tag{10.21}$$

where K is constant and D denotes the dimension of the fractal set (slope). The above algorithm has been applied to many patients and the dimension of their tumor cells have been found out. The above algorithm has been programmed and run in MATLAB.

Figure 10.4 is an example of Brain Tumor cells. Figure 10.5, Tumor cells in the other parts of the organ.

10.6 MASS RADIUS METHODS

The dimension of cluster liker items can be estimated with the help of the mass radius relation. The process involves choosing an origin point within the item, which is typically the center of mass, and calculating the number of particles (mass = pixels) at a radius r from the origin that comprise the object. The mass-radius relation for a

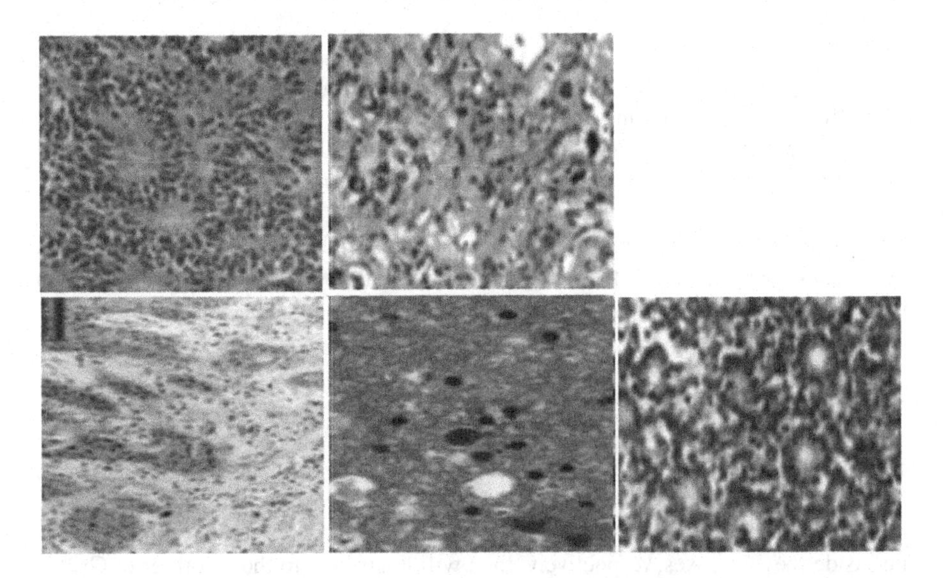

FIGURE 10.4 Brain tumor cells.

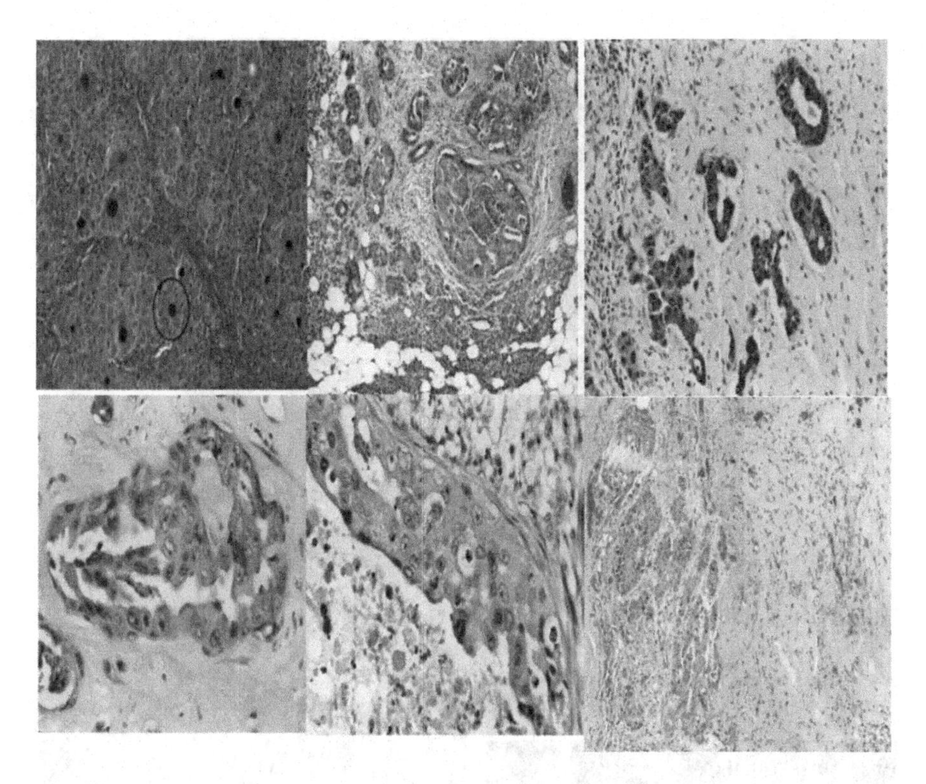

FIGURE 10.5 Tumor cells of the organ.

two-dimensional Euclidean object (a plane) is $M(r) \propto r^2$. The mass of a fractal object embedded in two dimensions changes with a fractional exponent, but the exponent is still the dimension as a result:

$$M(r) \infty \, r^D \tag{10.22}$$

$$D_{\text{mass-radius}} = \frac{\log M(r)}{\log(r)} \tag{10.23}$$

Then, compute the slope of the linear regression of $\log M(r)$ on $\log(r)$. There are two kinds of mistake in the graphical representation of the mass radius method dimensions in image analysis. Large estimations of areas at tiny radii are associated with the second, while the first is related to the estimation of the area of the circle scanned in a square matrix. The link between the area inside a certain radius and the size of this radius (or) box is defined by the mass dimension. This is carried out from different locations of origin and for different radii. The area's log-log plot as a function of radius can be used to estimate the mass dimension. One can compute the tumor's radius using this formula $r = \left(\frac{\text{Area}}{\pi}\right)^{1/2}$. We can get the radius for the various tumor cell types using this formula. The tumor grows in the organ according to the radius of the cells.

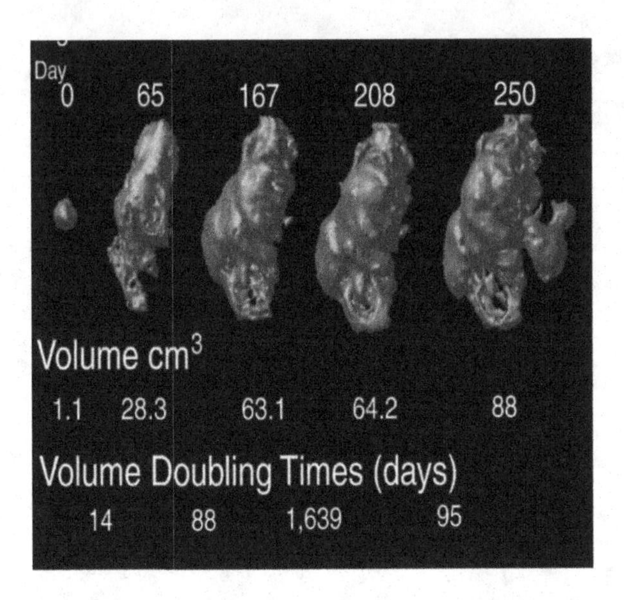

FIGURE 10.6 Mass of the tumor.

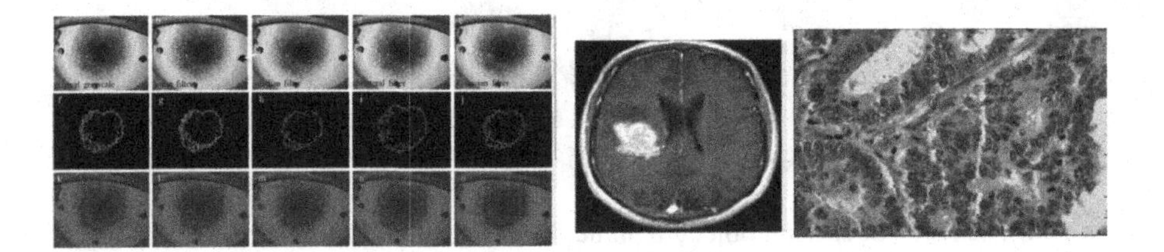

FIGURE 10.7 Radius of gyration.

Figure 10.6 evaluates the mass of the tumor, while Figure 10.7 evaluates the radius of gyration.

10.7 RADIUS OF GYRATION

The radius of gyration is a term used to describe several related measures of the size of an object, surface, or ensemble of points. It is calculated as the root mean square (rms) distance of the object's parts from either its center of gravity or an axis. The radius of gyration, Rg, of an object can also be used to estimate the fractal dimension, D. In growth processes, where the cluster size can be monitored as a function of time, Rg is calculated as

$$R_g = \sqrt{\frac{1}{N}\sum_{i=1}^{N} r_i^2} \tag{10.24}$$

where N is the number of particles in the cluster at a given time, and r_i is the distance from the ith particle to the center of mass of the cluster. If R_g can be calculated for several values of N and the object is self-similar, then the following relation allows for the computation of the fractal dimension $N \sim R_g^D$, $\log N = D \log R_g$,

$$D = \frac{\log(N)}{\log(R_g)} \tag{10.25}$$

The slope of the linear regression of log (N) on log (Rg) is D.

10.8 MODELS

This section explains a few models related to tumor formation. Tumors are aberrant cell growths that can be categorized into different models according to various standards. The following models are frequently employed in cancer research: By analyzing cells under a microscope, a tumor's histological appearance can be classified. Many forms, including squamous cell carcinoma, leukemia, sarcoma, lymphoma, and adenocarcinoma, are included in this categorization. The location of tumors throughout the body can also be used for classification, such as malignancies of the brain, lung, breast, and so forth. Thanks to developments in molecular biology, cancers can now be categorized according to molecular markers such as protein expression profiles, gene mutations, and patterns of gene expression. This classification aids in predicting prognosis and treatment response, for example, in cases of lung cancer with an EGFR mutation or HER2-positive breast cancer.

10.8.1 Eden Model

In a variation of the Eden concept, a lattice of immune and infectious individuals is initially formed, with a cell at its center. A location at the source is occupied, and then every vacant property nearby is recognized as a potential location for expansion. The next step involves randomly selecting and occupying one of these growth locations, each of which has an equal chance of being chosen. Both locations' unoccupied neighbors are then designated as expansion sites, and the process is repeated. The resultant aggregate is considered "compact," meaning that its fractal dimension, d, is equal to the Euclidean dimension of the embedding space. Furthermore, every location eventually becomes occupied. There are two types of geometry that can be used in all growth models: spherical geometry, which has a single seed at the origin, or strip geometry, which has seeds on a (d–1)–dimensional "plane" of size L^{d-1} and studies growth as a function of average height. This height is specified for the growth sides as

$$h = \frac{1}{N_s} \sum_{i=1}^{N_s} h_i \tag{10.26}$$

where Ns is the total number of growth sites, and hi is the distance of the ith growth site from the substrate. Defining the width of the growth zone via

$$\sigma(L,h) = \left[\frac{1}{N_s} \sum_{i=1}^{N_s} (h_i - h)^2 \right]^{\frac{1}{2}}, \tag{10.27}$$

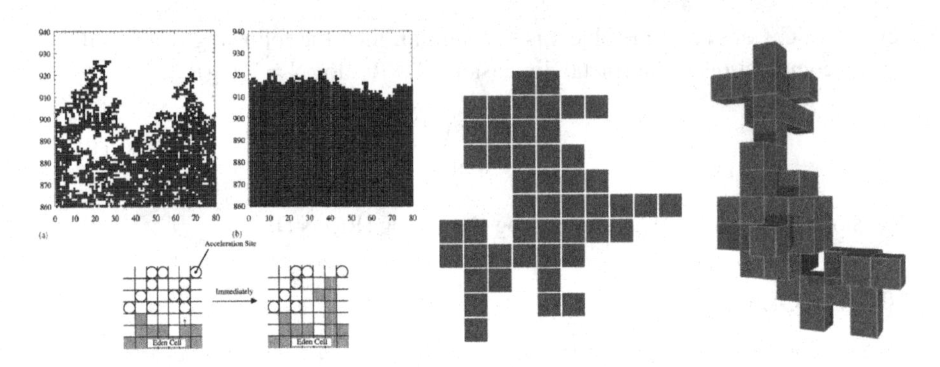

FIGURE 10.8 Growth of the tumor by the Eden model.

one can study the scaling of σ as a function of both L and h.

$$\sigma(L,h) = L^{\alpha} f\left(h/L^{z}\right), \tag{10.28}$$

$$\text{with} \quad f(x) \approx \begin{cases} x^{\beta} & \text{if } x \ll 1 \\ \text{const} & \text{if } x \gg 1. \end{cases} \tag{10.29}$$

Note that $x \ll 1$ means x is much less than 1 and $x \gg 1$ means x is much greater than 1. The fact that the diameters of the growth zones parallel and perpendicular to the substrate are not proportionate to one another is indicative of self-affinity $\alpha \neq 1$. This is the most basic concept for how a tumor might spread throughout an organ. These outcomes appear to be erratic. C++ can be used to program this. Figure 10.8 evaluate the growth of the Tumor by Eden Model.

10.8.2 Percolation Model

In the Eden model, each growth site has an equal chance of growing. Each growth site can be thought of as a cell for growth until it is eventually occupied. Models related to percolation alter both of these assumptions. In the conventional percolation model, every site on a lattice is either occupied (with probability p) or unoccupied (with probability (1 − p). As a result, each location can only be occupied once. If we begin at an occupied origin and proceed to its nearby sites, occupying a fraction p of them at random, this can be adapted into a growth model. Next, we examine all of the newly occupied sites' neighbors, disregarding those already designated as "unoccupied." After several iterations of this process, we obtain a distribution of clusters.

Figure 10.9 illustrates the percolation model.

The blood vessels within tumor development are believed to be randomly distributed. A randomly selected site is percolated using a random walker and is subsequently colored black. At each new time step, the vessel moves to a nearby fractal site, which also turns dark. The vessel computes the range at each time step t.

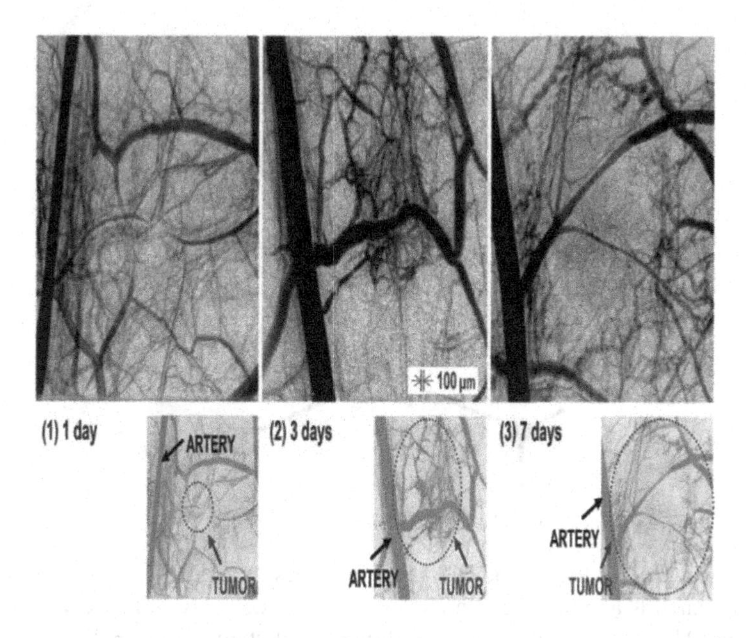

FIGURE 10.9 Percolation model.

$r \equiv \sqrt{\langle r^2 \rangle}$, root mean square from the local origin, where the blood vessel landed. The fractal dimension d_w determines how the time t scales with **r**: i.e., $r \sim t^{1/dw}$. Here we anticipate that d_w (the dimension of the blood vessel) is considerably larger than or equal to 2, so the blood vessel cannot reach many of the neighbors at each stage. It must therefore go back toward the original location. Therefore, r increases with t considerably more slowly than it would for an unconstrained random walk. Correlation length defines the linear size ξ of the finite clusters below and above p_c. The correlation length is defined as the mean distance between blood vessels on some finite cluster, when p approaches p_c, ξ increases as $\xi \sim (p - p_c)^{-v}$ with same exponent v below and above threshold. Below the percolation threshold concentration, all of these clusters are finite, with a size distribution function which behaves as $s^{-\tau}$for $s < \xi^{d_f}$ and decays exponentially for $s > \xi^{d_f}$. Where d_f means mass dimension with scales **r.**

10.8.3 Cayley Model

The Bethe lattice, commonly known as the Cayley tree, is a rigorous solution for the percolation issue. The blood flow that can percolate from one branch tree to another branch tree is known as the percolation threshold (p_c). The foundation of this idea is the mapping of each pore's threshold pressure to an occupation probability. One advantage of the Cayley tree is that the regime above can be explored, and the critical concentration is less than 1 and the regime above p_c can also be studied.

The Cayley tree is a non-looping structure that is created in the manner described below. We begin with a central site that is the source of z branches (of unit length). As new branches grow out from each side z–1, we gain z sites that make up the first cell

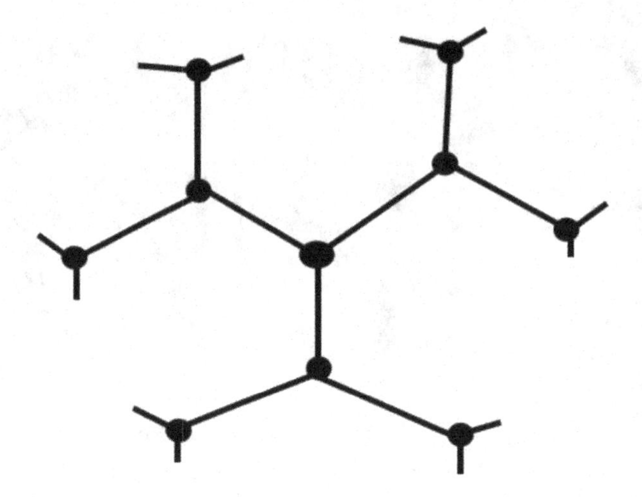

FIGURE 10.10 Two cells of a Cayley tree with z=3.

of the Cayley tree. In the second cell, these branches generate z(z−1) sides. At the end of each branch is another site. An endless Cayley tree is created by continuing this technique. The tree simplifies to a one-dimensional chain when z=2. Since there is just one connection connecting any two sites, the system is free of loops. The lattice is only described by the chemical distance l between two sites; the Euclidean distance r is meaningless in this context. For instance, there is precisely l chemical separation between the center site and a location of the lth cell. The Cayley tree can be thought of as an infinite-dimensional lattice since the exponential dependency can be interpreted as power-law behavior with an infinite exponent. We may anticipate that the critical exponents obtained for percolation on the Cayley tree will be the same as for percolation on any infinite-dimensional lattice based on the universality property.

Figure 10.10 evaluation of two cells of a Cayley tree with z=3.

The lth cell of the tree consists of $z(z-1)^{l-1}$ sites, increasing exponentially with l. In a d-dimensional Euclidean lattice, with d finite, the number of sites at distance l increases as l^{d-1}. The correlation function g(l), which is the mean number of sites on the same cluster at a distance l from an arbitrary occupied site, is where we start for the one-dimensional example. All l − 1 sites between two sites must be inhabited for them to be part of the same cluster if they are separated by a distance of l. Given that each cell has $z(z-1)^{l-1}$ sites and that the occupation is a random process with probability p, then follows that

$$g(l) = z(z-1)^{l-1} p^l$$

$$\equiv \frac{z}{z-1}\left[p(z-1)\right]^l \tag{10.30}$$

$$g(r) = 2p^r \tag{10.31}$$

For z=2, the Cayley tree reduces to a linear chain and Equation 10.28 reduces to Equation 10.29. From Equation 10.28 the critical concentration p_c can be easily derived. For l approaching infinity, the correlation function tends to zero exponentially for $p(z-1)<1$, and diverges for $p(z-1)>1$. Accordingly, an infinite cluster can be generated only if $p \geq 1/(z-1)$.

Hence,

$$p_c = \frac{1}{z-1} \tag{10.32}$$

The correlation length in l space $\xi_l^2 = \frac{\sum_{l=1}^{\infty} l^2 g(l)}{\sum_{l=1}^{\infty} g(l)}$

$$\xi_l^2 = p_c \frac{p_c + p}{(p_c - p)^2}, \quad p < p_c \tag{10.33}$$

As in $d = 1$, mass S of the finite clusters

$$S = 1 + \sum_{l=1}^{\infty} g(l) \tag{10.34}$$

Substituting Equation 10.28 into Equation 10.33, we find

$$S = p_c \frac{1+p}{p_c - p}, \quad p < p_c \tag{10.35}$$

which is a simple generalisation of the one-dimensional result.

Next we consider $n_s(p)$, the probability that a chosen site on the Cayley tree belongs to a cluster of s sites, divided by s. In the one-dimensional chain, n_s was simply the product of the probability p^s that s sites are occupied and the probability $(1-p)^2$ that the (two) perimeter sites are empty. For s cluster there exists only one realization. In general, the probability of finding a cluster with s sites having t perimeter sites is $p^s(1-p)^t$, and there exists more than one realization for a cluster of s sites. The general expression for n_s is therefore

$$n_s = \sum_t g_{s,t} p^s (1-p)^t. \tag{10.36}$$

On a Cayley tree, in contrast to the square lattice, there exists a unique relation between s and t. A cluster of one site is surrounded by z perimeter sites and a cluster of two sites has surrounded by $z + (z-2)$ perimeter sites. In general, a cluster of s sites has z–2 more perimeter sites than a cluster of s – 1 sites. Denoting the number of perimeter sites of s cluster by $t(s)$, we obtain

$$t(s) = z + (s-1)(z-2) \tag{10.37}$$

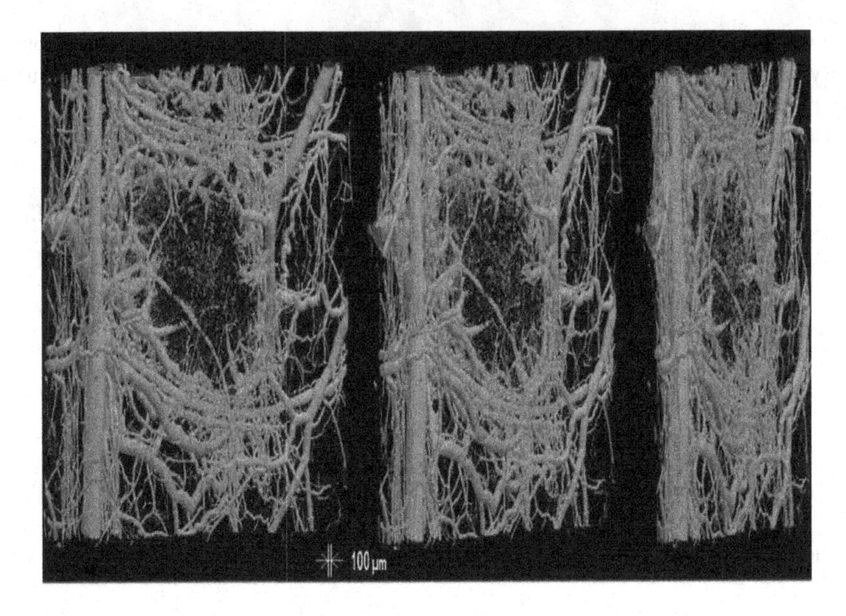

FIGURE 10.11 Growth of blood vessel-like Cayley tree.

Then for the Cayley tree, Equation 10.35 reduces to

$$\begin{aligned} n_s(p) &= g_s p^s (1-p)^{2+(z-2)s} \\ &\equiv g_s (1-p)^2 \left[p(1-p)^{z-2}\right]^s, \end{aligned} \tag{10.38}$$

where g_s is simply the number of configurations for a s-site cluster.

Note that Equation 10.37 is general and holds for all values of p. Figure 10.11 evaluates the growth of blood vessel-like Cayley trees on different days.

The modeling of the Cayley tree is based on the blood vessel system of the tumor. The shortest path of blood flow is modeled using Kruskal's Algorithm, which is programmed in C++ (Appendix A). The growth of the blood vessel can be explained by the Cayley tree, as articulated by the following proposition.

Let $E^{(*)}$ be the set of infinite Cayley tree. Let A be a subset (finite or infinite) of the infinite Cayley tree $E^{(*)}$. A node $\gamma \in E^{(*)}$ is a lower bound of the set A iff $\gamma \leq \alpha$ for all $\alpha \in A$. (when we think of branches, we might say that γ is a "common prefix" for the set A.) A mode β is a greatest lower bound for the set A iff β is a lower bound for A and $\gamma \leq \beta$ for any other lower bound γ for A. If α and β are two blood vessels, then we may form a blood vessel $\alpha\beta$ called concatenation, by listing the symbols of the blood vessel α followed by the symbols of the blood vessel β.

10.9 RESULTS

The results of Box Counting Method of different images of the tumor are presented in Tables 10.1 and 10.2. In order to use the algorithm I, let r be the box (mesh) size in

TABLE 10.1
Data Analysis of Different Brain Tumors Using the Box Counting Method

Scaling	1	D_B	2	D_B	3	D_B	4	D_B	5	D_B	6	D_B	7	D_B	8	D_B	9	D_B	10	D_B
2	1148		376		863		1975.5		3294.5		2748.5		3906		3326		6301		4993.5	
3	687.33		294.67		583.55		1108.668		1889.6674		1844.6674		2428		2042		3088.1094		2725.8867	
4	440		235.75		397		695.720		1265.25		1328		1649		1372.5		1868		1763.1250	
5	300		194		285		467.4		938.0011		996.1995		1195		1037.2002		1233.9199		1206.7588	
6	209	1.7	158.67	1.02	206.56	1.5	333.1667	1.8	700.8336	1.4	773.5003	1.3	878	1.5	772.6664	1.5	901.7774	1.8	907.5560	1.6
7	159.97		126.14		157.83		246.1836		568.4286		605.2861		678.1428		599.1426		676.1429		694.4290	
8	124		105.25		121		190.7813		456.2344		484.25		531.7969		474.25		530.25		536.3438	
9	96.65		89.33		97.54		150.7407		388.336		394.3331		429.0125		388.2839		421.5306		433.0985	
10	80		72		80		122.1		325		323.3999		353.25		327		347.08		353.3397	

TABLE 10.2
Data Analysis of Other Parts of the Organ Using the Box Counting Method

Scaling	1	D_B	2	D_B	3	D_B	4	D_B	5	D_B	6	D_B	7	D_B	8	D_B
2	605		841.5		3316		674		22,935		10,890		20,844		26,624	
3	402.8892		756.6666		2083.6670		420		13,295.70		6018.33		11,650.68		13,163.31	
4	325.7		700.25		1513.5		345		8052.5		4020.5		7312.5		7,758	
5	271.59994		661.6002		1209.2		266.8		5245.98		2948.6		4962.98		5102.4	
6	226.8890	1.02	633.8334	1.23	989.6668	.5	208.78	1.67	3656.32	1.8	2290.66	1.94	3560.66	1.7	3597.32	1.9
7	209.3470		589.5715		854.7145		176.55		2690.43		1892.85		2652.57		2664.57	
8	192.6875		558.6250		745.75		145.5		2059.62		1559.25		2043.37		2049.37	
9	173.3580		516.0001		671.4445		132.23		1628.14		1343.11		1622.14		1624.48	
10	168.2000		525.1002		6145.3001		103.2		1318.4		1159.3		1316.8		1315.8	

pixels and N(r) be the number of boxes that cover objects (cells). Using linear regression method we find the lines at fit the points of $\ln N(r)$ verses $\ln(1/r)$. The fractal dimension is nothing but the slope of the line.

The radius of the cell and the expansion of blood vessels determine the tumor's bulk. As the radius of gyration increases over time, the tumor's mass radius indicates that the self-similarity characterizing cell proliferation in the tumor is reflected by both the mass radius method and the radius of gyration. In fact, scaling features are believed to have evolved from the microvascular structure responsible for delivering nutrients to the cells of the organism. The Eden model can describe the tumor's tissue dissemination. Additionally, the Percolation model can simulate blood flow within a tumor. Blood flows through vessels that supply tumor-forming cells in the targeted organ or brain with essential nutrients. The Cayley tree can represent the growth of blood vessels, which is explained by Kruskal's Algorithm. This provides insight into the invasiveness of the growth [11].

Figure 10.12 represents the brain tumor, while Figure 10.13 graphically depicts other parts of the organ.

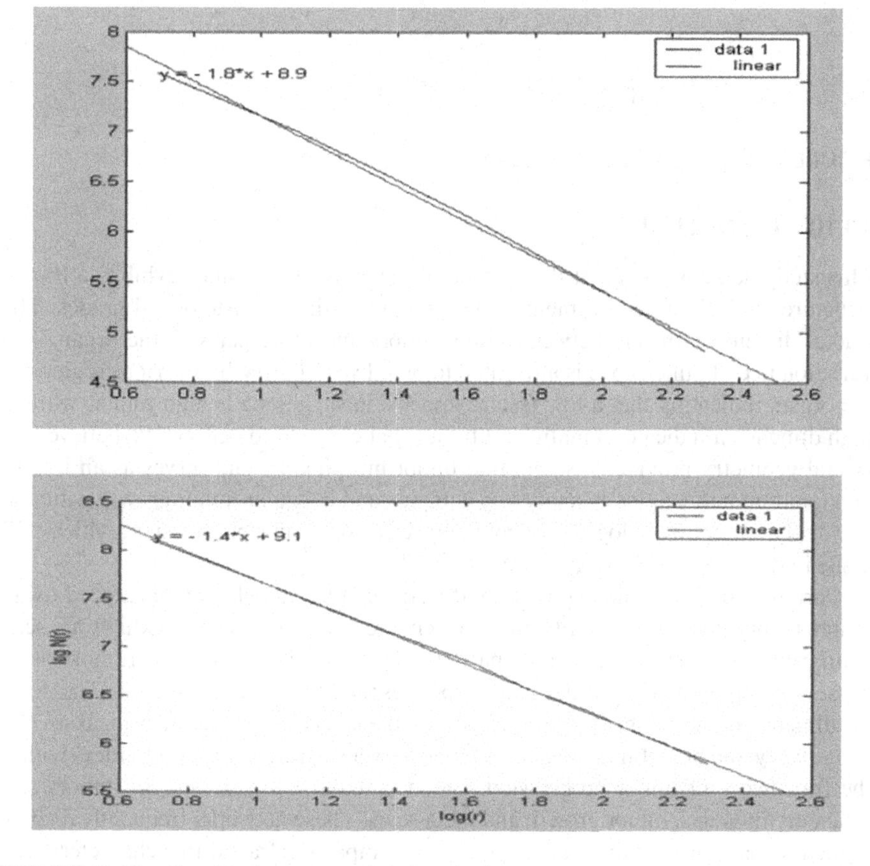

FIGURE 10.12 Graphical representation for a brain tumor.

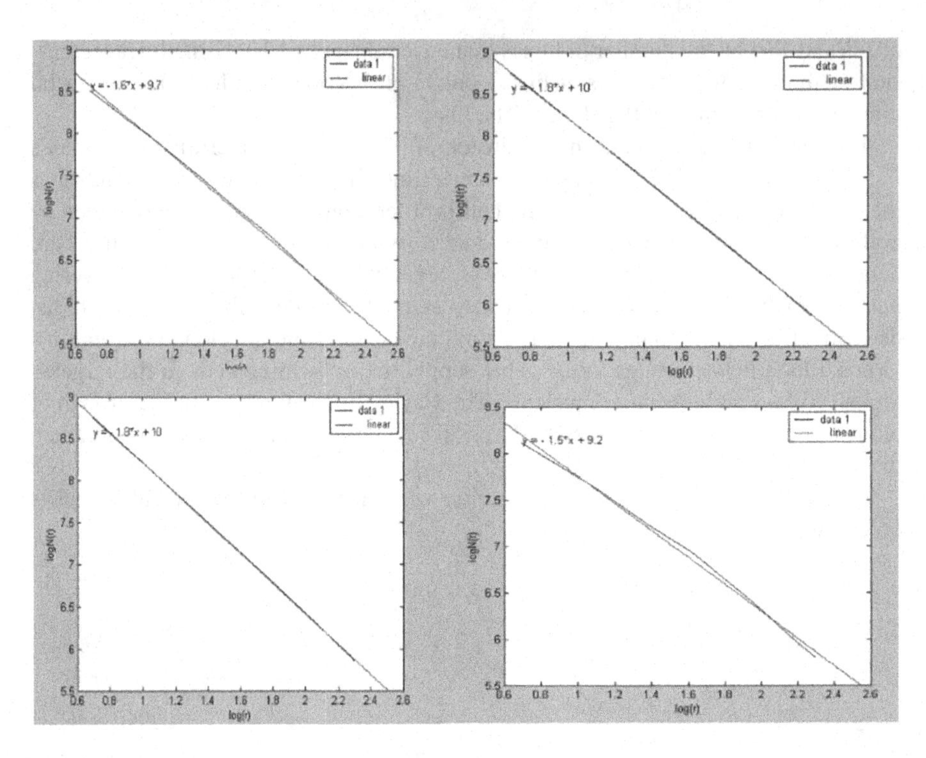

FIGURE 10.13 Graphical representation of an organ.

10.10 CONCLUSION

This study demonstrates that the surface of the cells in the tumor exhibits a fractal structure, with the fractal dimension varying according to histological grades. The fractal dimension differs between brain tumors and other parts of the organ. The variation in cell dimensions is attributed to blood vessel growth in various regions of the organ, indicating that a low fractal dimension suggests a benign tumor, while a high dimension indicates a malignant tumor that can lead to cancer. We believe that fractal geometry provides insights into tumor morphology and serves as an important tool for mathematically analyzing complex and irregular tumor growth patterns. Our report offers pathologists a new perspective on tumor progression, which will assist in determining treatment strategies [12].

Certain aspects of tumor growth and shape can be modeled and analyzed using graph theory and fractals. Fractals, which are geometric forms exhibiting self-similarity at various scales, can be particularly useful in this context. Tumors often possess complex and asymmetrical shapes, especially when malignant. The fractal dimension can quantify the intricacy of these shapes. It can also measure the irregularity and branching patterns of blood vessel networks or tumor borders within the framework of tumor progression. Blood vascular networks within tumors significantly influence tumor growth and metastasis. These networks frequently display branching patterns that recur at various scales, imparting fractal-like characteristics. The structure of these vascular networks within tumors can be modeled and analyzed using graph theory, where nodes represent vessel junctions and edges signify vessel segments. Fractal analysis can then describe the efficiency and complexity of these networks. The tumor microenvironment is highly heterogeneous and spatially structured, consisting of the extracellular matrix, various cell types, and signalling chemicals. Fractal analysis can measure the spatial distribution and arrangement of different components within this microenvironment. This study can be further enhanced by employing graph theory, which models the connections and interactions between various components as a network.

REFERENCES

1. Huang, K., and Yan, H., Offline signature verification based on geometric feature extraction and neural network classification, *Pattern Recognition*, Vol. 30(1), pp. 9–17 (1997).
2. Cohn, D., Optimal systems I: The vascular system, *Bulletin of Mathematical Biophysics*, Vol. 16(1), pp. 59–74 (1954).
3. Wagner, C. D., and Persson, P. B., Chaos in the cardiovascular system: An update, *Cardiovascular Research*, Vol. 40(2), pp. 257–264 (1998).
4. Mandelbrot, B. B., *The Fractal Geometry of Nature*, W.H. Freeman, San Francisco, CA, p. 460 (1982).
5. Hastings, H. M., and Sugihara, G., Fractals, *A User's Guide for the Natural Sciences*, Oxford University Press, New York (1993).
6. Hutchinson, J., Fractals and self-similarity, *Indiana University Mathematics Journal*, Vol. 30(5), pp. 713–747 (1981).
7. Mandlebrot, B. B., How long is the coast of Britain? Statistical self similarity and fractional dimension, *Science*, Vol. 156, pp. 636–638 (1967).
8. Caldwell, C. B., Moran, E. L., and Bogoch, E. R., Fractal dimension as a measure of altered trabecular bone in experimental inflammatory arthritis. *Journal of Bone Minerals Research*, Vol. 13(6), pp. 978–985 (1998).
9. Ott, E., *Chaos*, Cambridge University Press, New York (1988).
10. Baish, J. W., and Jain, R. K., Fractals and cancer, *Cancer Research*, Vol. 60, pp. 3683–3688 (2000).
11. Fierraz, J., Ortega-Garcia, J., Ramos, D., and Rodriguez, J. G., HMM based on-line signature verification: Feature extraction and signature modeling, *Pattern Recognition Letters*, Vol. 28(16), pp. 2325–2334 (2007).
12. Zook, J. M., and Iftekharuddin, K. M., Statistical analysis of fractal-based brain tumor detection algorithms, *Magnetic Resonance Imaging*, Vol. 23(5), 671–678 (2005).

11 Fractals in Circuit Theory

11.1 INTRODUCTION

In circuit theory, fractals have been applied in fascinating ways that provide new methods for creating and evaluating electrical circuits. One of the main fields influenced by fractals is the creation of fractal antennas. The self-similarity of fractals allows these antennas to achieve wideband and multiband properties. Conventional antennas are typically designed to operate within a narrow frequency range. In contrast, fractal antennas can function simultaneously across multiple frequencies due to their self-similar structure. This characteristic makes them highly adaptable and suitable for a wide range of communication systems, such as cellular phones, wireless networks, and radar systems. Furthermore, fractal geometries enable antennas to be made smaller without compromising performance. By repeatedly reducing a fractal antenna structure, designers can produce compact antennas that perform well over a broad frequency range. This miniaturization is particularly beneficial for portable electronic devices with limited space. Additionally, fractal-based circuit components such as capacitors and inductors are also under study. The design of these components incorporates fractal geometry, resulting in improved miniaturization, reduced parasitic effects, and increased self-resonant frequency, among other enhanced electrical features. Fractal-based passive components may offer significant advantages in integrated circuits, miniature electronic systems, and high-frequency applications. Fractal-based methods have also been employed in the design of filters, transmission lines, and other circuit elements. Fractal shapes can enhance electromagnetic interference suppression, impedance matching, and signal propagation, thereby improving the overall reliability and performance of electronic circuits. In conclusion, fractals present intriguing opportunities for the development of compact, wideband antennas, miniaturized circuit components, and optimized transmission lines within the realm of circuit theory. As technology advances, we can anticipate further research and development aimed at utilizing fractal geometry to enhance the performance and efficiency of electrical devices. Fractals, invented by mathematicians, serve as powerful tools for measuring complex objects, solving computational problems, and uncovering patterns in various systems. People who are fractals are enthralled with pictures of themselves that resemble them. The two sources of the fractals can be traced back to people who measured the length and boundary of the British coast. Additionally, the weather patterns of the world exhibit fractal characteristics. Fractal geometry is described as "the repetition of similar patterns over a wide range of length scales." Nature displays similar space-filling structures in brain networks, vascular systems, and cloud formations. However, not all materials exhibiting self-similarity can be classified as fractals. Fractals also utilize the sine function through four distinct fixed-point iterations.

DOI: 10.1201/9781003481096-11

11.2 MULTISIM LABORATORY

This chapter explains how to generate a sine wave in the Multisim 14.2 laboratory using virtual simulation software [1]. It produces simulation results that accurately resemble a sine wave. Multisim is implemented as a core course in electronics. Fractal-generating programs have been evaluated based on the self-similarity of fractal images. Multisim utilizes iterated function systems governed by fixed geometrical replacement rules. Fractal patterns have been extensively designed, often within a scale dimension, due to the feasible limits of physical time and space. This software is used in digital signal processing, linking modeling and testing with theoretical concepts. It presents a method for clipping in the digital domain, focusing on the main component parameters of an analog sine wave signal. The output is tabulated as time sequences with corresponding voltage values. Lagrange Interpolation evaluates the polynomial equation f(x), calculated based on the data points, to find the function's expression for time sequences, even when the arguments are equally spaced.

This work presents an investigation of a sine fractal wave produced by NI Multisim using numerical methods. It is a computer program that contains a schematic or model of an electronic circuit. Software algorithms for modeling methods are widely available, and this program is used to forecast and assess the potential success of the electronic circuit. The sine function has only one true fixed point, which is zero. Thus, the sine wave can be expressed as an infinite series, allowing its range to include complex, positive, and negative values. The discussion includes Poincaré's notions for extensions, which contrast with Weierstrass's synopsis.

Self-similar images are called fractals, which represent broken or subdivided objects. In 1918, mathematician Felix Hausdorff introduced the concept of complex geometric shapes that commonly exhibit fractional dimensions. An exact self-similarity fractal satisfies self-similarity conditions very strongly. Scale invariance is a principle that measures length, energy, etc. It represents the proper form of self-similarity at any magnitude value, with a small part of the object resembling the whole. This is defined by an iterative function, demonstrating accurate self-similarity.

1. The commands for all functions are seen in the Menu bar.
2. The Design Toolbox allows you to explore a schematic hierarchy, show or hide different layers, and manage various file types in a project.
3. Use the buttons on the component toolbar to select components from the Multisim Database. The buttons on the standard toolbar can save, print, cut, and paste data.
4. The buttons in the View toolbar enable you to change the display of the screen.
5. The simulation toolbar contains buttons to initiate, halt, and perform additional simulation operations. Common Multisim function buttons are located on the main toolbar.
6. The In Use List contains a record of all the parts utilized in the design
7. Each instrument's button is located in the Instruments toolbar.
8. Use the left-to-right scroll to facilitate handling larger drawings.
9. Build your circuit in the circuit window. The active circuit is shown in the active lab.

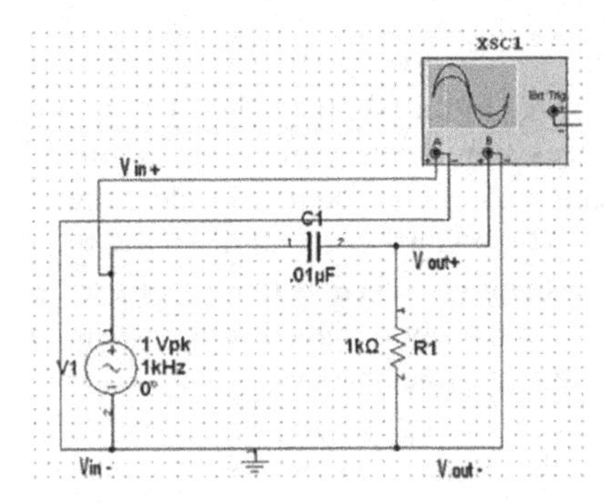

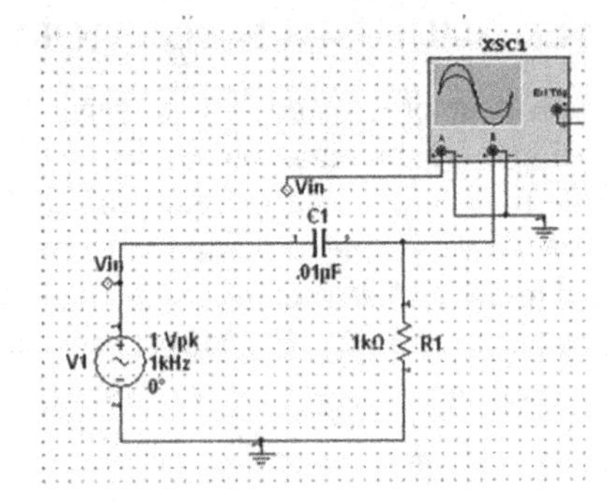

FIGURE 11.1 Multisim laboratory.

11.3 RELATED WORKS

A fractal has a distinctive appearance; it is a well-crafted framework with random proportions. The Hausdorff dimension measures the roughness or complexity of fractal graphs. It is superior to one-dimensional measurements and has a clear, recurrent definition. However, not all self-similar structures are considered fractals. For example, Euclidean shapes have an accurately similar appearance but lack the characteristics of fractals, making it sufficient to describe them in Euclidean terms. A fractal appears identical at various scales, with each part being precisely similar [3]. It can be represented through a computerized program that models the circuit pattern of an electronic circuit. Many software algorithms are available for simulation techniques.

These software algorithms are used to forecast and assess the performance of electronic circuits. After designing a circuit, it is essential to test its functionality and validate its operation. If necessary, changes can then be made to enhance its efficiency. This electronic design software includes extensive component libraries, offering a variety of useful components such as multimeters, oscilloscopes, signal generators, logic converters, and analyzers.

The software allows for testing the functionality of a circuit without physically building it. It can simulate both analog and digital circuits, making it more efficient and less time-consuming to design complex circuits. This approach saves time and money while reducing hardware resource wastage. Once satisfactory results are achieved, the hardware can be built to match the accurate results from the simulation. The circuit can also be redesigned based on the simulation outcomes. At each level of design and simulation, performance can be verified and compared with theoretical responses.

Figure 11.2 shows the Multisim manual modeling.

11.4 CIRCUIT SIMULATION

Circuit simulation is a computer application that utilizes a model or circuit pattern of an electronic circuit. Several software methods are available for simulation techniques,

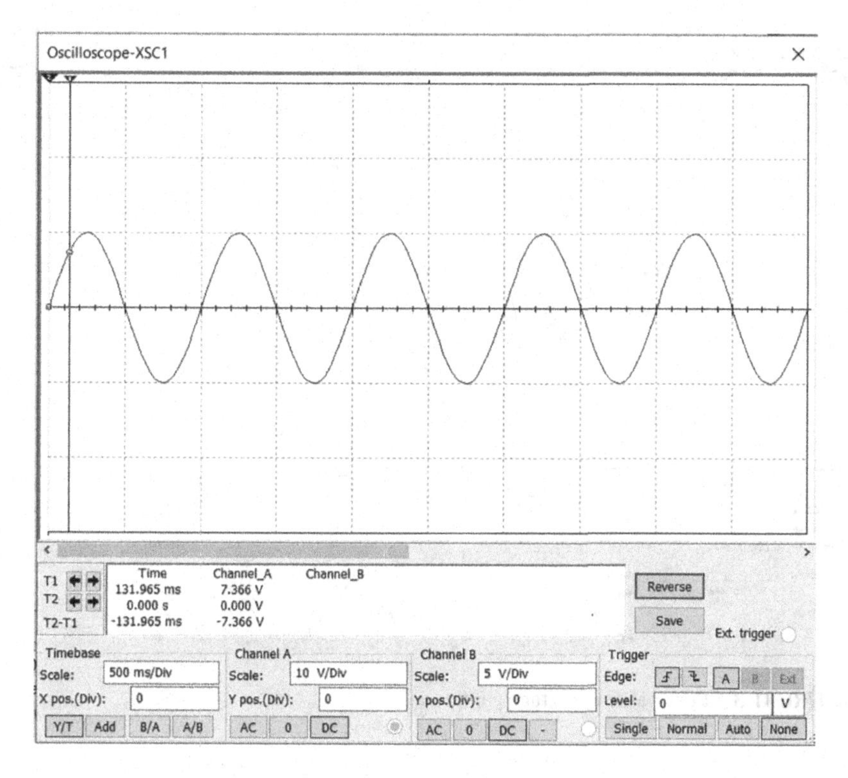

FIGURE 11.2 Modeling.

which are used to develop and examine the circuit. The behavior and performance of the circuit are predicted and established using this software. Designing electrical and electronic circuits is often a costly, time-consuming, and technically complex process. Engineers seek to verify the circuit's functionality after it has been designed and make any necessary adjustments to optimize its efficiency. To start, open Multisim by clicking Start → Programs → National Instruments → Circuit Design Suite 11.0 → Multisim 11.0. Create a new file by selecting File → New → Design.

Electrical and electronic applications have led to the development of increasingly intricate, costly, and time-consuming circuits. The goal is to validate the engineers' designs and test the circuit's functionality after completion. If needed, adjustments can be made to enhance the circuit's efficiency. This section focuses on several typical applications of the virtual simulation tool Multisim in analog electronic technology education, improving the quality of live, productive classroom instruction. Multisim includes additional component libraries intended for electronic design, featuring numerous useful components such as oscilloscopes, multimeters, logic converters, analyzers, and signal generators [4]. It is used to simulate both analog and digital circuitry. The value of y is calculated for the corresponding value of x, where $X_0 < X_i < X_n$. If the value of f(x) = y is to be determined at some point y in the interval $[x_0, x_n]$ and y is not one of the tabulated points, the value of y is estimated by using the known values of f(x) at the surrounding points. Interpolation is the process of finding the most appropriate estimate for the missing data. Data have been listed as (Table 11.1)

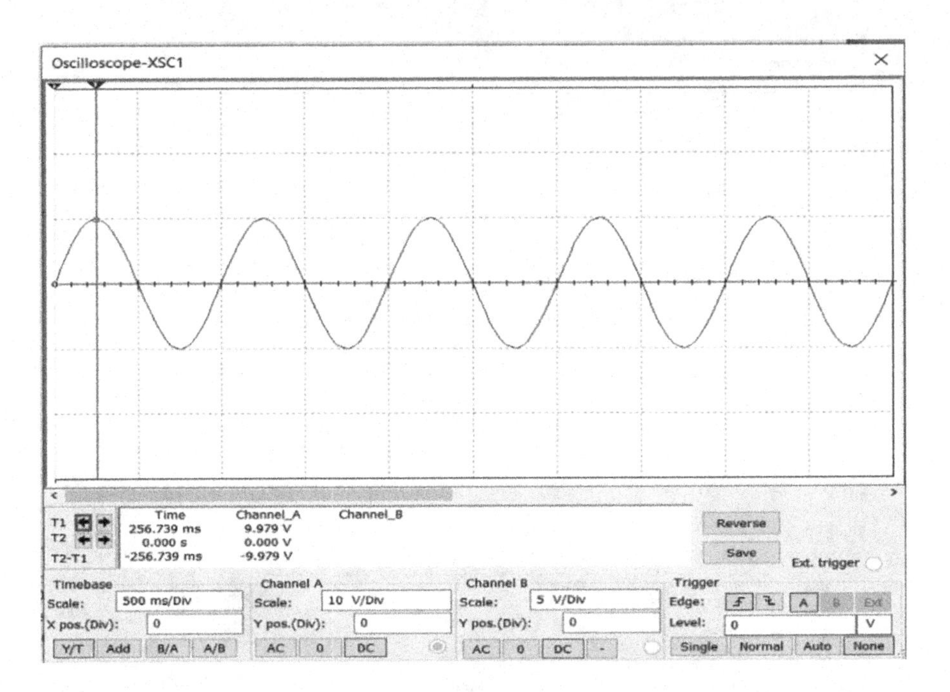

FIGURE 11.3 Peak value of voltage captured.

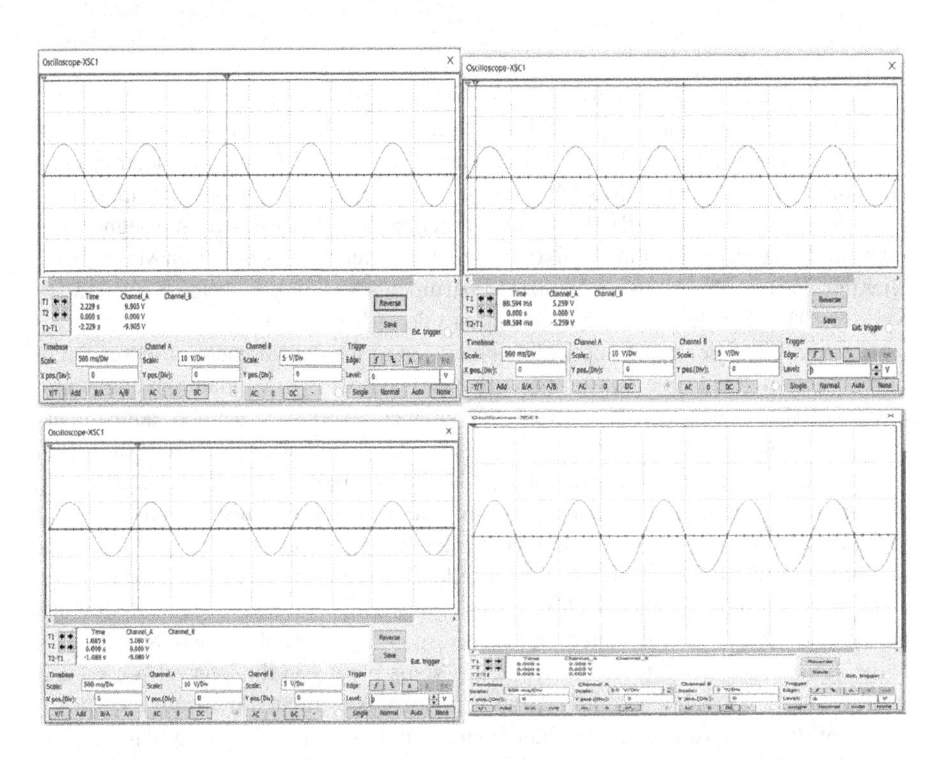

FIGURE 11.4 Output categorized into self-similarity fractals.

TABLE 11.1
Output of Voltage Obtained by Simulation Software

S. No	Time Taken (ms)	Voltage (V)
1	0	0
2	44.929	2.777
3	102.696	6.013
4	131.965	7.366
5	243.902	9.945
6	256.739	9.979

TABLE 11.2
Data Taken from Decreasing Function

S. No	Time	Voltage
1	306.093	18.642
2	372.977	14.332
3	423.256	9.242
4	462.791	4.611
5	483.372	1.455
6	500.000	−0.0002236

11.4.1 Observation

At the beginning, the output of the wave function is a continuously increasing function. The sine wave reaches its maximum voltage value of 9.979 at t = 256.739 ms. After reaching this peak, it suddenly begins to decrease. The sine wave drops to zero and then goes to a negative value of −0.0002236 at t = 500 ms. It continues to decrease as it approaches the peak value of the negative output.

Table 11.2 has decreasing function of sine wave

11.4.2 Time Sequence and Voltage

Time x	44.929 ms	102.696 ms	263.158 ms
Voltage y	2.777 V	6.013 V	9.943 V

Apply Lagrange interpolation to find the voltage corresponding to x = 131.965

Let us consider

$X_0 = 44.929$ ms,

$X_1 = 102.696$ ms,

$X_2 = 263.158$ ms,

$Y_0 = 2.777$ mV,

$Y_1 = 6.013\,mV$,
$Y_2 = 9.943\,mV$

Lagrangeinterpolation formula

$$Y = F(131.965)$$

$$= \frac{(X - X_1)(X - X_2)}{(X_0 - X)(X_0 - X_2)} Y_0 + \frac{(X - X_0)(X - X_2)}{(X_1 - X_0)(X_1 - X_2)} Y_1 + \frac{(X - X_0)(X - X_1)}{(X_2 - X_0)(X_2 - X_1)} Y_2$$

$$= \frac{(131.965 - 102.696)(131.965 - 263.158)}{(44.929 - 102.696)(44.929 - 263.158)}(2.777)$$

$$+ \frac{(131.965 - 44.929)(131.965 - 263.158)}{(102.696 - 44.929)(102.696 - 263.158)}(6.013)$$

$$+ \frac{(131.965 - 44.929)(131.965 - 102.696)}{(263.158 - 44.929)(263.158 - 102.696)}(9.943)$$

$$= 7.28457 \cong 7.3(\text{Approximately})$$

Let's consider the below from the output of the signal

Time x	306.093 ms	462.791 ms	500.000
Voltage y	18.642 mV	4.611 mV	–0.0002236 mV

Let consider the following:
$X_0 = 306.093$ ms,
$X_1 = 462.791$ ms,
$X_2 = 500.00$ ms,
$Y_0 = 18.642\,V$,
$Y_1 = 4.611\,V$,
$Y_2 = -0.0002236\,V$
Find the voltage corresponding to time $t = 423.256$.
Lagrange interpolation formula:

$$Y = F(X)$$

$$= \frac{(X - X_1)(X - X_2)}{(X_0 - X)(X_0 - X_2)} Y_0 + \frac{(X - X_0)(X - X_2)}{(X_1 - X_0)(X_1 - X_2)} Y_1 + \frac{(X - X_0)(X - X_1)}{(X_2 - X_0)(X_2 - X_1)} Y_2$$

$$= \frac{(423.256 - 462.791)(423.256 - 500.000)}{(306.093 - 462.791)(306.093 - 500.000)}(18.642)$$

$$+ \frac{(423.256 - 306.093)(423.256 - 500.000)}{(462.791 - 306.093)(462.791 - 500.000)}(4.611)$$

$$+ \frac{(423.256 - 306.093)(423.256 - 462.791)}{(500.000 - 306.093)(500.000 - 462.791)}(-0.0002236)$$

$$= 8.97 \approx 9(\text{Approximately})$$

Peak value of voltage through the increasing function f(x) = 19.936 at t = 258.140. After this point, the function will begin to decrease. There is a common difference between the time-frequency and voltage. A constant ratio of increasing time frequency and decreasing voltage has been established. Exact self-similarity has been created through the signal output.

11.5 ANALYSIS OF FRACTALS THROUGH NI SIMULATION SOFTWARE

A fractal is a self-similar structure, and all periodic functions are considered fractals [5]. This has been analyzed using data obtained from Multisim Simulation [6]. The model figure shows that the software output remains a sine wave, which is a periodic function. The output has been estimated for voltage corresponding to the time sequence. The full periodic wave has been categorized into four sections, starting at zero voltage at time t = 0 ms. Approximately every 250 ms, the wave is moderated. It reaches a peak value, which can be either maximum or minimum, at the end of 250 ms.

A sine wave is a geometric waveform that oscillates, moving up, down, or side to side periodically. It is defined by f(x) = sin x and has a well-defined "s" shape, which is the graphical representation of a general function. A sinusoidal wave is a curve defined in terms of the sine trigonometric function. An original sinusoidal wave has a single frequency without harmonics or additional frequencies. It illustrates how a variable's amplitude varies with time. An alternating current or voltage within a wire produces a simple oscillating wave, regarded as a sine wave. Frequency measures the amount of time required for a sine wave to complete one cycle in one second, with hertz being the unit of measurement. The angle of the sine wave varies from −1 to +1. When a voltage is applied through a resistor, current flows, which is captured through an oscilloscope. Analyzing the time sequence with voltage shows an initially fractal-increasing function. After reaching the peak voltage, it decreases in correlation with the specific time sequence and voltage.

Various methods, such as electrochemical patterns, digital photographs, harmonic sounds, and rhymes, have been employed to produce model fractals. Cisco modeling involves creating patterns in a real three-dimensional environment. Recurring fractal models are typically developed using fractal creation software. The fractal-increasing function is derived from the voltage-based time sequence analysis. The Multisim simulation program features transparent confluence, an intuitive interface, and graphical representation, serving as the platform for the simulation laboratory [7]. Numerous empirical applications in electrical circuit design systems, prototyping, and testing have emerged from this lab. The teaching and learning process in the electrical and electronic engineering laboratory has undergone a significant transformation with the introduction of this simulation lab, which is utilized to understand theoretical data and solve problems related to pure theory through numerical analysis (Tables 11.3 and 11.4).

TABLE 11.3
Analysis of Time Sequence and Voltage from the Output of NI Software between Cycle I and Cycle II

Time (ms)	Time Difference	Voltage (mV)	Difference of Voltage	Time (s)	Time Difference (s)	Voltage (mV)	Difference of Voltage
First Quadrant							
Cycle I				Cycle II			
5.071		0.637126		1.004		0.510734	
10.142	5.071	1.274	0.636874	1.009	0.005	1.146	0.635266
15.213	5.071	1.908	0.634	1.014	0.005	1.775	0.629
20.284	5.071	2.542	0.634	1.019	0.005	2.403	0.628
25.355	5.071	3.167	0.625	1.024	0.005	3.032	0.629
30.426	5.071	3.793	0.626	1.029	0.005	3.661	0.629
35.497	5.071	4.418	0.625	1.034	0.005	4.29	0.629
40.568	5.071	5.041	0.623	1.04	0.006	4.918	0.628
45.639	5.071	5.638	0.597	1.045	0.005	5.519	0.601
50.71	5.071	6.236	0.598	1.05	0.005	6.117	0.598
55.781	5.071	6.834	0.598	1.055	0.005	6.714	0.597
60.852	5.071	7.432	0.598	1.06	0.005	7.312	0.598
65.923	5.071	8.03	0.598	1.065	0.005	7.91	0.598
70.994	5.071	8.627	0.597	1.07	0.005	8.508	0.598
76.065	5.071	9.174	0.547	1.075	0.005	9.065	0.557
81.136	5.071	9.718	0.544	1.08	0.005	9.609	0.544
86.207	5.071	10.261	0.543	1.085	0.005	10.153	0.544
91.278	5.071	10.805	0.544	1.09	0.005	10.696	0.543
96.349	5.071	11.349	0.544	1.095	0.005	11.24	0.544
101.42	5.071	11.893	0.544	1.1	0.005	11.784	0.544
106.491	5.071	12.377	0.484	1.105	0.005	12.284	0.5
111.562	5.071	12.846	0.469	1.111	0.006	12.753	0.469
116.633	5.071	13.315	0.469	1.116	0.005	13.221	0.468
121.704	5.071	13.784	0.469	1.121	0.005	13.69	0.469
126.775	5.071	14.253	0.469	1.126	0.005	14.159	0.469
131.846	5.071	14.722	0.469	1.131	0.005	14.628	0.469
136.917	5.071	15.133	0.411	1.136	0.005	15.057	0.429
141.988	5.071	15.509	0.376	1.141	0.005	15.433	0.376
147.059	5.071	15.885	0.376	1.146	0.005	15.809	0.376
152.13	5.071	16.261	0.376	1.151	0.005	16.185	0.376
157.201	5.071	16.637	0.376	1.156	0.005	16.561	0.376
162.272	5.071	17.013	0.376	1.161	0.005	16.937	0.376
167.343	5.071	17.339	0.326	1.166	0.005	17.285	0.348
172.414	5.071	17.608	0.269	1.171	0.005	17.554	0.269
177.485	5.071	17.876	0.268	1.176	0.005	17.822	0.268
182.556	5.071	18.145	0.269	1.182	0.006	18.091	0.269
187.627	5.071	18.413	0.268	1.187	0.005	18.36	0.269

(Continued)

TABLE 11.3 (*Continued*)
Analysis of Time Sequence and Voltage from the Output of NI Software between Cycle I and Cycle II

Time (ms)	Time Difference	Voltage (mV)	Difference of Voltage	Time (s)	Time Difference (s)	Voltage (mV)	Difference of Voltage
192.698	5.071	18.652	0.239	1.192	0.005	18.628	0.268
197.769	5.071	18.915	0.263	1.192	0	18.885	0.257
202.84	5.071	19.066	0.151	1.202	0.01	19.036	0.151
207.911	5.071	19.217	0.151	1.207	0.005	19.187	0.151
212.982	5.071	19.368	0.151	1.212	0.005	19.338	0.151
218.053	5.071	19.519	0.151	1.217	0.005	19.489	0.151
223.124	5.071	19.67	0.151	1.222	0.005	19.64	0.151
228.195	5.071	19.821	0.151	1.227	0.005	19.791	0.151
233.266	5.071	19.832	0.011	1.232	0.005	19.826	0.035
238.337	5.071	19.859	0.027	1.237	0.005	19.854	0.028
243.408	5.071	19.887	0.028	1.242	0.005	19.881	0.027
248.479	5.071	19.914	0.027	1.247	0.005	19.909	0.028
253.55	5.071	19.942	0.028	1.253	0.006	19.936	0.027
258.621	5.071	19.969	0.027	1.258	0.005	19.964	0.028
Second Quadrant							
Cycle I				Cycle II			
263.692	5.071	19.875	−0.094	1.263	0.005	19.895	−0.069
268.763	5.071	19.778	−0.097	1.268	0.005	19.798	−0.097
273.834	5.071	19.681	−0.097	1.273	0.005	19.7	−0.098
278.905	5.071	19.584	−0.097	1.278	0.005	19.603	−0.097
283.976	5.071	19.487	−0.097	1.283	0.005	19.506	−0.097
294.118	10.142	19.195	−0.292	1.288	0.005	19.409	−0.097
299.189	5.071	18.977	−0.218	1.293	0.005	19.238	−0.171
304.26	5.071	18.759	−0.218	1.298	0.005	19.02	−0.218
309.331	5.071	18.541	−0.218	1.303	0.005	18.803	−0.217
314.402	5.071	18.323	−0.218	1.313	0.01	18.585	−0.218
319.473	5.071	18.105	−0.218	1.324	0.011	18.367	−0.218
324.544	5.071	17.815	−0.29	1.329	0.005	18.149	−0.218
329.615	5.071	17.484	−0.331	1.334	0.005	17.881	−0.268
334.686	5.071	17.154	−0.33	1.339	0.005	17.55	−0.331
339.757	5.071	16.824	−0.33	1.344	0.005	17.22	−0.33
344.828	5.071	16.493	−0.331	1.349	0.005	16.89	−0.33
349.899	5.071	16.163	−0.33	1.354	0.005	16.559	−0.331
354.97	5.071	15.784	−0.379	1.359	0.005	16.229	−0.33
360.041	5.071	15.354	−0.43	1.364	0.005	15.87	−0.359
365.112	5.071	14.924	−0.43	1.369	0.005	15.44	−0.43
370.183	5.071	14.494	−0.43	1.374	0.005	15.01	−0.43
375.254	5.071	14.064	−0.43	1.379	0.005	14.58	−0.43
380.325	5.071	13.364	−0.7	1.384	0.005	14.15	−0.43

(Continued)

TABLE 11.3 (*Continued*)
Analysis of Time Sequence and Voltage from the Output of NI Software between Cycle I and Cycle II

Time (ms)	Time Difference	Voltage (mV)	Difference of Voltage	Time (s)	Time Difference (s)	Voltage (mV)	Difference of Voltage
385.396	5.071	13.177	−0.187	1.389	0.005	13.72	−0.43
390.467	5.071	12.664	−0.513	1.395	0.006	13.28	−0.44
395.538	5.071	12.151	−0.513	1.4	0.005	11.74	−1.54
400.609	5.071	11.637	−0.514	1.405	0.005	11.227	−0.513
405.68	5.071	11.124	−0.513	1.41	0.005	10.714	−0.513
410.751	5.071	10.611	−0.513	1.415	0.005	10.2	−0.514
415.822	5.071	10.087	−0.524	1.42	0.005	9.626	−0.574
420.892	5.07	9.511	−0.576	1.425	0.005	9.049	−0.577
425.963	5.071	8.934	−0.577	1.43	0.005	8.473	−0.576
431.034	5.071	8.357	−0.577	1.435	0.005	7.896	−0.577
436.105	5.071	7.78	−0.577	1.44	0.005	7.319	−0.577
441.176	5.071	7.204	−0.576	1.445	0.005	6.742	−0.577
446.247	5.071	6.627	−0.577	1.45	0.005	6.133	−0.609
451.318	5.071	6.009	−0.618	1.455	0.005	5.515	−0.618
456.389	5.071	5.391	−0.618	1.46	0.005	4.897	−0.618
461.46	5.071	4.773	−0.618	1.466	0.006	4.278	−0.619
466.531	5.071	4.155	−0.618	1.471	0.005	3.66	−0.618
471.602	5.071	3.537	−0.618	1.476	0.005	3.042	−0.618
476.673	5.071	2.919	−0.618	1.481	0.005	2.413	−0.629
481.744	5.071	2.286	−0.633	1.486	0.005	1.778	−0.635
486.815	5.071	1.65	−0.636	1.491	0.005	1.142	−0.636
491.886	5.071	1.015	−0.635	1.496	0.005	0.506236	−0.63576
496.957	5.071	0.379138	−0.63586	1.501	0.005	−0.12936	−0.6356
502.028	5.071	−0.256485	−0.63562				
			Third Quadrant				
	Cycle I				Cycle II		
512.17	10.142	−1.523	−1.26652	1.511	0.005	−1.397	−0.63202
517.241	5.071	−2.152	−0.629	1.516	0.005	−2.026	−0.629
522.312	5.071	−2.781	−0.629	1.521	0.005	−2.655	−0.629
527.383	5.071	−3.409	−0.628	1.526	0.005	−3.284	−0.629
532.454	5.071	−4.038	−0.629	1.53	0.004	−3.75	−0.466
537.525	5.071	−4.667	−0.629	1.534	0.004	−4.17	−0.42
542.596	5.071	−5.28	−0.613	1.537	0.003	−4.589	−0.419
547.667	5.071	−5.878	−0.598	1.54	0.003	−5.007	−0.418
552.738	5.071	−6.475	−0.597	1.544	0.004	−5.406	−0.399
557.809	5.071	−7.073	−1.195	1.547	0.003	−5.805	−0.399
562.88	5.071	−7.671	−0.598	1.551	0.004	−6.302	−0.497
567.951	5.071	−8.269	−0.598	1.556	0.005	−6.819	−0.517

(*Continued*)

TABLE 11.3 (*Continued*)
Analysis of Time Sequence and Voltage from the Output of NI Software between Cycle I and Cycle II

Time (ms)	Time Difference	Voltage (mV)	Difference of Voltage	Time (s)	Time Difference (s)	Voltage (mV)	Difference of Voltage
573.022	5.071	−8.847	−0.578	1.56	0.004	−7.335	−0.516
578.093	5.071	−9.391	−0.544	1.566	0.006	−8.014	−0.679
583.164	5.071	−9.935	−0.544	1.575	0.009	−9.013	−0.999
588.235	5.071	−10.479	−0.544	1.579	0.004	−9.483	−0.47
593.306	5.071	−11.023	−0.544	1.583	0.004	−9.953	−0.47
598.377	5.071	−11.566	−0.543	1.588	0.005	−10.423	−0.47
603.448	5.071	−12.096	−0.53	1.592	0.004	−10.894	−0.471
608.518	5.07	−12.565	−0.469	1.596	0.004	−11.364	−0.47
613.59	5.072	−13.034	−0.469	1.601	0.005	−11.834	−0.47
618.661	5.071	−13.503	−0.469	1.605	0.004	−12.264	−0.43
623.732	5.071	−13.972	−0.469	1.61	0.005	−12.669	−0.405
628.803	5.071	−14.44	−0.468	1.614	0.004	−13.075	−0.406
633.874	5.071	−14.907	−0.467	1.618	0.004	−13.481	−0.406
638.945	5.071	−15.283	−0.376	1.623	0.005	−13.886	−0.405
644.016	5.071	−15.659	−0.376	1.627	0.004	−14.292	−0.406
649.087	5.071	−16.035	−0.376	1.632	0.005	−14.697	−0.405
654.158	5.071	−16.411	−0.376	1.636	0.004	−15.062	−0.365
659.229	5.071	−16.787	−0.376	1.64	0.004	−15.387	−0.325
664.3	5.071	−17.163	−0.376	1.645	0.005	−15.712	−0.325
669.371	5.071	−17.446	−0.283	1.649	0.004	−16.038	−0.326
674.442	5.071	−17.715	−0.269	1.654	0.005	−16.363	−0.325
679.513	5.071	−17.984	−0.269	1.658	0.004	−16.688	−0.325
684.584	5.071	−18.252	−0.268	1.662	0.004	−17.013	−0.325
689.655	5.071	−18.521	−0.269	1.667	0.005	−17.303	−0.29
694.726	5.071	−18.79	−0.269	1.671	0.004	−17.535	−0.232
699.797	5.071	−18.976	−0.186	1.675	0.004	−17.768	−0.233
704.868	5.071	−19.127	−0.151	1.68	0.005	−18	−0.232
709.939	5.071	−19.278	−0.151	1.684	0.004	−18.232	−0.232
715.01	5.071	−19.429	−0.151	1.689	0.005	−18.465	−0.233
720.081	5.071	−19.58	−0.151	1.693	0.004	−18.697	−0.232
725.152	5.071	−19.731	−0.151	1.697	0.004	−18.904	−0.207
730.223	5.071	−19.815	−0.084	1.702	0.005	−19.034	−0.13
735.294	5.071	−19.843	−0.028	1.706	0.004	−19.165	−0.131
740.365	5.071	−19.87	−0.027	1.711	0.005	−19.295	−0.13
745.436	5.071	−19.898	−0.028	1.715	0.004	−19.426	−0.131
750.507	5.071	−19.925	−0.027	1.719	0.004	−19.556	−0.13
755.578	5.071	−19.953	−0.028	1.724	0.005	−19.687	−0.131
760.649	5.071	−19.933	0.02	1.728	0.004	−19.804	−0.117
				1.732	0.004	−19.827	−0.023

(*Continued*)

TABLE 11.3 (*Continued*)
Analysis of Time Sequence and Voltage from the Output of NI Software between Cycle I and Cycle II

Time (ms)	Time Difference	Voltage (mV)	Difference of Voltage	Time (s)	Time Difference (s)	Voltage (mV)	Difference of Voltage
				1.737	0.005	−19.851	−0.024
				1.741	0.004	−19.875	−0.024
				1.746	0.005	−19.899	−0.024
				1.75	0.004	−19.922	−0.023
				1.754	0.004	−19.946	−0.024
				1.759	0.005	−19.969	−0.023
Fourth Quadrant							
Cycle I				**Cycle II**			
765.72	5.071	−19.836	0.097	1.768	0.009	−19.801	0.168
770.791	5.071	−19.739	0.097	1.772	0.004	−19.718	0.083
775.862	5.071	−19.642	0.097	1.776	0.004	−19.634	0.084
780.933	5.071	−19.545	0.097	1.781	0.005	−19.55	0.084
786.004	5.071	−19.448	0.097	1.785	0.004	−19.466	0.084
791.075	5.071	−19.325	0.123	1.789	0.004	−19.382	0.084
796.146	5.071	−19.108	0.217	1.794	0.005	−19.206	0.176
801.217	5.071	−18.89	0.218	1.798	0.004	−19.017	0.189
806.288	5.071	−18.672	0.218	1.803	0.005	−18.829	0.188
811.359	5.071	−18.454	0.218	1.807	0.004	−18.641	0.188
816.43	5.071	−18.236	0.218	1.811	0.004	−18.452	0.189
821.501	5.071	−18.013	0.223	1.816	0.005	−18.264	0.188
826.572	5.071	−17.682	0.331	1.82	0.004	−18.075	0.189
831.643	5.071	−17.352	0.33	1.825	0.005	−17.813	0.262
836.714	5.071	−17.022	0.33	1.829	0.004	−17.528	0.285
841.785	5.071	−16.692	0.33	1.833	0.004	−17.242	0.286
846.856	5.071	−16.631	0.061	1.838	0.005	−16.956	0.286
851.927	5.071	−16.031	0.6	1.842	0.004	−16.671	0.285
856.998	5.071	−15.612	0.419	1.846	0.004	−16.385	0.286
862.069	5.071	−15.182	0.43	1.851	0.005	−16.099	0.286
867.14	5.071	−14.752	0.43	1.855	0.004	−15.759	0.34
872.311	5.171	−14.322	0.43	1.86	0.005	−15.387	0.372
877.282	4.971	−13.892	0.43	1.864	0.004	−15.016	0.371
882.353	5.071	−13.462	0.43	1.868	0.004	−14.644	0.372
887.424	5.071	−12.972	0.49	1.873	0.005	−14.272	0.372
892.495	5.071	−12.459	0.513	1.877	0.004	−13.9	0.372
897.566	5.071	−11.945	0.514	1.882	0.005	−13.528	0.372
902.637	5.071	−11.432	0.513	1.886	0.004	−13.119	0.409
907.708	5.071	−10.919	0.513	1.89	0.004	−12.676	0.443
912.779	5.071	−10.406	0.513	1.895	0.005	−12.232	0.444

(*Continued*)

TABLE 11.3 (*Continued*)
Analysis of Time Sequence and Voltage from the Output of NI Software between Cycle I and Cycle II

Time (ms)	Time Difference	Voltage (mV)	Difference of Voltage	Time (s)	Time Difference (s)	Voltage (mV)	Difference of Voltage
917.85	5.071	−9.857	0.549	1.899	0.004	−11.788	0.444
922.921	5.071	−9.28	0.577	1.904	0.005	−11.344	0.444
927.992	5.071	−8.703	0.577	1.908	0.004	−10.9	0.444
933.063	5.071	−8.127	0.576	1.912	0.004	−10.456	0.444
938.134	5.071	−7.55	0.577	1.917	0.005	−9.991	0.465
943.205	5.071	−6.973	0.577	1.921	0.004	−9.492	0.499
948.276	5.071	−6.38	0.593	1.925	0.004	−8.994	0.498
953.347	5.071	−5.762	0.618	1.93	0.005	−8.495	0.499
958.418	5.071	−5.144	0.618	1.934	0.004	−7.996	0.499
963.489	5.071	−4.526	0.618	1.939	0.005	−7.497	0.499
968.56	5.071	−3.908	0.618	1.943	0.004	−6.998	0.499
973.631	5.071	−3.29	0.618	1.9475	0.0045	−6.49	0.508
978.702	5.071	−2.667	0.623	1.952	0.0045	−5.956	0.534
983.773	5.071	−2.032	0.635	1.956	0.004	−5.421	0.535
988.844	5.071	−1.396	0.636	1.961	0.005	−4.887	0.534
993.915	5.071	−0.760512	0.635488	1.965	0.004	−4.352	0.535
998.986	5.071	−0.124889	0.635623	1.969	0.004	−3.818	0.534
1004	5.014	0.510734	0.635623	1.974	0.005	−3.283	0.535
				1.978	0.004	−2.747	0.536
				1.982	0.004	−2.197	0.55
				1.987	0.005	−1.647	0.55
				1.991	0.004	−1.097	0.55
				1.996	0.005	−0.54752	0.549477
				2	0.004	0.002236	0.549759

TABLE 11.4
Analysis of Time Sequence and Voltage from the Output of NI Software between Cycle I and Cycle II

Time (ms)	Time Difference	Voltage (mV)	Difference of Voltage	Time (s)	Time Difference (s)	Voltage (mV)	Difference of Voltage
			First Quadrant				
Cycle I				Cycle II			
5.071		0.637126		1.004		0.510734	
10.142	5.071	1.274	0.636874	1.009	0.005	1.146	0.635266
15.213	5.071	1.908	0.634	1.014	0.005	1.775	0.629
20.284	5.071	2.542	0.634	1.019	0.005	2.403	0.628

(*Continued*)

TABLE 11.4 (*Continued*)
Analysis of Time Sequence and Voltage from the Output of NI Software between Cycle I and Cycle II

Time (ms)	Time Difference	Voltage (mV)	Difference of Voltage	Time (s)	Time Difference (s)	Voltage (mV)	Difference of Voltage
25.355	5.071	3.167	0.625	1.024	0.005	3.032	0.629
248.479	5.071	19.914	0.027	1.247	0.005	19.909	0.028
258.621	5.071	19.969	0.027	1.258	0.005	19.964	0.028
Second Quadrant							
Cycle I				Cycle II			
263.692	5.071	19.875	−0.094	1.263	0.005	19.895	−0.069
268.763	5.071	19.778	−0.097	1.268	0.005	19.798	−0.097
273.834	5.071	19.681	−0.097	1.273	0.005	19.7	−0.098
278.905	5.071	19.584	−0.097	1.278	0.005	19.603	−0.097
283.976	5.071	19.487	−0.097	1.283	0.005	19.506	−0.097
294.118	10.142	19.195	−0.292	1.288	0.005	19.409	−0.097
299.189-	5.071	18.977	−0.218	1.293	0.005	19.238	−0.171
496.957	5.071	0.379138	−0.63586	1.496	0.005	0.506236	−0.63576
502.028	5.071	0.256485	−0.63562	1.501	0.005	−0.12936	−0.6356
Third Quadrant							
Cycle I				Cycle II			
512.17	10.142	−1.523	−1.26652	1.511	0.005	−1.397	−0.63202
517.241	5.071	−2.152	−0.629	1.516	0.005	−2.026	−0.629
522.312	5.071	−2.781	−0.629	1.521	0.005	−2.655	−0.629
527.383	5.071	−3.409	−0.628	1.526	0.005	−3.284	−0.629
532.454	5.071	−4.038	−0.629	1.53	0.004	−3.75	−0.466
537.525	5.071	−4.667	−0.629	1.534	0.004	−4.17	−0.42
542.596	5.071	−5.28	−0.613	1.537	0.003	−4.589	−0.419
684.584	5.071	−18.252	−0.268	1.662	0.004	−17.013	−0.325
750.507	5.071	−19.925-	−0.027	1.75	0.004	−19.922	−0.023
755.578	5.071	−19.953	−0.028	1.754	0.004	−19.946	−0.024
760.649	5.071	−19.933	0.02	1.759	0.005	−19.969	−0.023
Fourth Quadrant							
Cycle 1				Cycle 2			
765.72	5.071	−19.836	0.097	1.768	0.009	−19.801	0.168
770.791	5.071	−19.739	0.097	1.772	0.004	−19.718	0.083
775.862	5.071	−19.642	0.097	1.776	0.004	−19.634	0.084
780.933	5.071	−19.545	0.097	1.781	0.005	−19.55	0.084
988.844	5.071	−1.396	0.636	1.961	0.005	−4.887	0.534
993.915	5.071	0.76051	0.63548	1.965	0.004	−4.352	0.535
998.986	5.071	0.12488	0.63562	1.996	0.005	−0.54752	0.549477
1004	5.014	0.510734	0.635623	2	0.004	0.002236	0.549759

11.6 RESULT

From the above table in the first quadrant of Cycle I through Channel A, the time duration has increased to 5.071 ms at each level of increment. Correspondingly, the voltage has increased within a specific range of sequences. The voltage starts at zero and reaches a peak value of 19.969 at t = 258.621. In the second quadrant, the peak value of voltage suddenly decreases in a particular manner, reaching zero at the end of the second quadrant. After this, the voltage decreases continuously, attaining a minimum value of zero before reaching a negative value in the third quadrant.

The minimum voltage recorded is −19.933 mV at time t = 760.649. From Cycle III, the wave moves upward, and the voltage gradually increases with a minimal level of difference. The negative voltage level increases slightly until it reaches zero at a specific point in time.

These four quadrant values should be consistent across all cycles. The second cycle's voltage data are similar to those of the first cycle, with both exhibiting the same level of voltage frequency and time sequence. The iteration method is the most effective technique for identifying self-similar fractal graphs. By using this method, the cycles of the periodic function project the same value, which indicates that the output voltage remains a periodic function.

11.7 APPLICATION OF MULTISIM IN RADIO FREQUENCY

The application of Multisim in the realm of radio frequency (RF) circuits is significant, offering engineers and designers a powerful tool for simulation and analysis [2]. Here's a brief overview of how Multisim is used in the context of RF circuits. RF Circuit Simulation: Multisim allows engineers to simulate RF circuits, including components such as antennas, filters, amplifiers, and mixers. This capability is crucial during the design phase of RF systems, as it helps to understand circuit behavior. RF engineering is a field of electronics that heavily utilizes Multisim, a potent simulation program. The following are some typical uses for Multisim in RF design: Circuit Creation: Multisim provides an intuitive framework for creating and simulating RF circuits, including mixers, amplifiers, filters, oscillators, and more. The graphical schematic editor enables engineers to build RF circuits and model their behavior under various operational scenarios. Antenna Simulation: Engineers can use Multisim to design and evaluate various types of antennas, such as patch, fractal, and dipole antennas. Antenna performance factors, including radiation pattern, gain, bandwidth, and impedance matching, can be simulated effectively. Transmission Line Modeling: Transmission lines are essential components of RF systems for transferring RF signals with minimal loss and distortion. Multisim allows engineers to model and simulate different types of transmission lines, including coaxial cables, microstrip lines, and striplines, enabling examination of their impedance matching, signal propagation, and impedance characteristics. Filter Design and Optimization: Multisim simplifies the design and optimization of RF filters, including band-pass, band-stop, high-pass, and low-pass filters. Engineers can simulate the frequency response, insertion loss, and return loss of the filters to ensure they meet the necessary requirements for signal conditioning and interference suppression. Amplifier Analysis: Engineers can construct and analyze RF amplifiers using Multisim, including popular configurations like class-A, class-B, class-C, and class-AB amplifiers. Users can simulate performance aspects, such as distortion characteristics, noise figure, output power, and gain. Mixer and Modulator Design: RF mixers and modulators, which are employed in frequency conversion and modulation/demodulation processes, can be designed and simulated with Multisim. Engineers can examine converter gain, linearity, modulation quality, and mixer spurious products to ensure proper operation in RF communication systems. Signal Examination Tools: Multisim offers tools to analyze RF signal properties, such as frequency spectrum, time-domain waveform, and frequency-domain response. By examining signal integrity, distortion, noise, and other performance indicators, engineers can troubleshoot RF circuits effectively. Overall, Multisim is an invaluable tool for RF engineers, allowing them to build, simulate, and evaluate a variety of RF circuits and systems. This capability facilitates quicker RF design validation, optimization, and prototyping prior to hardware implementation.

11.8 CONCLUSION

The modeled fractal has been developed using images, patterns, rhymes, harmonic sounds, and more. Analyzing the time sequence with voltage reveals a fractal-increasing function. Multisim simulation software offers a transparent, user-friendly interface with graphical representation, serving as an effective simulation laboratory platform. This laboratory has facilitated numerous empirical applications in design systems, prototyping, and testing of electrical circuits. It serves as a milestone for enhancing the teaching and learning process within the electrical and electronic engineering laboratory. The platform aids in understanding theoretical concepts and overcoming pure theoretical challenges through numerical analysis. The oscilloscope displays a sine wave, and the output generated from the modeling process via Multisim software can produce highly stimulating harmonious distributions, which serve as criteria for fractal analysis or for observation purposes. The sine wave test pattern is a convenient function related to frequency. However, the simple formula for calculating the sine wave feedback aspect can present challenges during experimentation. By utilizing numerical analysis, the convenience of the sine wave feedback aspect in calculating system performance is demonstrated.

REFERENCES

1. Coltman, J. W., The specification of imaging properties by response to a sine wave input, *JOSA*, Vol. 44(6), pp. 468–471 (1954).
2. Corbin, M. J., and Butler, G. F., MulTiSIM: An object-based distributed framework for mission simulation, *Simulation Practice and Theory*, Vol. 3(6), pp. 383–399 (1996).
3. Falconer, K. J., *Techniques in Fractal Geometry*, Wiley, Chichester, p. 3 (1997).
4. Mahata, S., Maiti, A., and Maiti, C. K., Cost-effectiveweb-based electronics laboratory using NI MultiSim, LabVIEW and ELVIS II, In *2010 International Conference on Technology for Education*, pp. 242–243, IEEE (2010). https://doi.org/10.1109/T4E.2010.5550110.

5. Mandelbrot, B. B., Fractal geometry: What is it, and what does it do? *Proceedings of the Royal Society of London A: Mathematical and Physical Science*, Vol. 143(1864), pp. 3–16 (1989).
6. Xian, F., and Lai, X. Z., Based on multisim simulation digital logic experiment teaching reform. *Research and Exploration in Laboratory*, Vol. 38(9), pp. 228–232, 297 (2019).
7. Zhengdong, X. L., et al., Application of multisim simulation software in teaching of analog electronic technology, *Journal of Physics: Conference Series*, Vol. 1544(1), p. 012063 (2020). https://doi.org/10.1088/1742-6596/1544/1/012063.

12 Fractals in Architecture

12.1 INTRODUCTION

The study of fractals and their attributes in relation to graphs is known as fractal graph theory. Fractals are geometric shapes characterized by self-similarity at various scales, meaning that individual parts of the object resemble the whole. In contrast, graph theory examines graphs, which are mathematical structures made up of vertices connected by edges [1]. The scaling characteristics and complexity of these graphs are measured by their fractal dimension. Fractal graphs exhibit self-similarity, where the overall structure of the graph is mirrored in its smaller sections. Scholars develop techniques and algorithms to create graphs that resemble fractals, aiming to produce graphs with specific fractal characteristics. Applications of fractal graph theory are found across various disciplines, including computer science, physics, biology, and geography. For instance, it can be used to model intricate networks such as biological systems, the internet, or the composition of natural landscapes

12.2 APPLICATION OF FRACTALS IN ARCHITECTURE

However, a new trend has emerged that produces complex images that cannot be easily defined. This type of architecture is referred to as fractal building. Similar to the significance of metallic ratios in the past, the study of modern fractal dimensions in architecture has gained importance [2]. Through fractal analysis, the composition of a building's elevation can be evaluated in a mathematically quantifiable manner using the box counting approach. The goal of this chapter is to illuminate how fractal geometry influences the perception of buildings in relation to their surroundings and their role in aesthetics. Fractals with a high degree of density do not always possess an unpleasant visual quality, and their appeal is not necessarily related to their fractal dimensions.

12.3 FRACTAL ARCHITECTURE

African architecture, design, and art feature a plethora of examples of fractal models. These expressions highlight the social, religious, and cultural makeup of settlements. Examples of African architecture display fractal characteristics as a result of the settlement's organizational and structural features. Large villages were generally preferred over cities in European politics. The European Architecture Fractal Theorem encompasses both contemporary art and Da Vinci's sketches, reflecting the continent's political systems, socialism, and culture. For instance, the 20th century saw the rise of Russian and Soviet art and architecture under Kazimir Malevich. There is a growing interest in 3D architectural projects and sculptures from this period. Remarkable examples of fractals in architecture can be found in the field

DOI: 10.1201/9781003481096-12

of Arkhitektonics [3]. The main part of a building is often filled with a cascade of smaller replicas, maintaining a roughly satisfactory scale range, despite the tendency for these structures to obscure scale and minimize the gap between buildings and people. India is known for its architecturally significant structures, particularly its arch temples, whose construction remains a mystery to this day. Numerous temples and monuments in India and Southeast Asia incorporate fractal architecture, characterized by towers encircled by smaller towers, with the only differences among them being their ratios.

12.4 FRACTALS IN NATURE

Trees have thousands of branches, and forests have thousands of trees. Hindu architecture reflects the fractal nature of Hindu cosmology. However, there is no surprise regarding buildings in Euclidean forms. Fractals found in architecture are derived from nature and exhibit patterns that mimic the natural world [4]. This fractal architecture has emerged in different cultures and can be divided into three categories. Typically, fractals are used separately for water and sewage systems that serve a group of buildings or a city. Plumbing systems have been developed in urban areas; for example, the Romans built complex aqueducts in Europe to supply potable water to their cities, many of which are still visible today. However, human waste and water-filled ditch systems were removed from cities using carts or buckets, which were then dumped outside or into lakes or streams.

12.5 RANDOM FRACTAL GRAPHS

Analyzing the characteristics of graphs produced by stochastic processes is essential for studying random fractal graphs. These graphs incorporate randomness and exhibit fractal-like properties. Fractal graph theorists employ methods from graph theory, topology, geometry, and fractal geometry to investigate the complex properties of these intricate structures. As scientists and mathematicians explore new applications and expand their understanding of fractal graphs and their characteristics, the field continues to develop. While fractal graphs, as mathematical structures, had no direct bearing on the governance of kings in ancient India, considering the broader notion of fractals and self-similarity may lead to metaphorical or analogical comparisons with specific facets of ancient Indian social structures and government.

12.5.1 Hierarchy and Self-Similarity

In ancient Indian society, the king or ruler typically occupied the top position in the social hierarchy, with tiers of nobility, administration, and commoners descending from them. The smaller social units within this hierarchical structure resembled the larger governing structure, reflecting a fractal pattern. Just as fractals display self-similarity at various scales, the power of the king was mirrored in the roles of village chiefs, local officials, and regional governors, each overseeing smaller regions [5].

12.5.2 Dharma and Order

The dharma (duty, righteousness) principles governed ancient Indian rulers. The preservation of justice, peace, and order in society was a key component of the dharma concept. Dharma can be thought of metaphorically as a guiding principle, similar to the self-similar patterns seen in fractals. A commitment to dharma helped maintain stability and coherence in governance at different levels of society, much like fractals display repeating patterns that preserve structure and coherence. Decentralized Governance: While central authority was common in ancient Indian kingdoms, there was also some decentralization of governance. Local leaders, such as chieftains or landlords, had a great deal of autonomy within their territories. The self-similar properties of fractals, in which smaller components reflect the larger structure, can be likened to this decentralized governance. Just as smaller fractal shapes fit into a larger geometric pattern, each local ruler functioned within the broader framework of the kingdom.

12.5.3 Social Organization

Indian society's stratification comprised social classes called varnas and occupational groups called jatis, each with specific roles and responsibilities. Self-similar patterns can be observed in this social structure, where the various jatis and communities reflect the overarching varna system. The varna-jati system encapsulated a complex social structure with interrelated components, much like fractals that display intricate patterns at different scales.

These connections illustrate how mathematical concepts like fractals can enhance our understanding of historical and cultural phenomena, even if they are conceptual and metaphorical rather than strictly mathematical.. A comparison can be drawn between the hierarchical organization of King Ashoka's empire in ancient India and the fractal graph concept of self-similarity.

12.6 ASHOKA'S REIGN IS NOTABLE FOR SEVERAL REASONS

12.6.1 Military Conquests

From roughly 268 to 232 BCE, Ashoka, also referred to as Ashoka the Great, was an ancient Indian emperor who ruled the Maurya Empire. He is regarded as one of the most important kings in Indian history. The son of Emperor Bindusara and the grandson of Chandragupta Maurya, who established the Mauryan Empire, Ashoka's reign is noteworthy for several reasons.

12.6.2 Conversion to Buddhism

After the bloody Kalinga War, which claimed many lives, Ashoka underwent a dramatic change of heart. Moved by the misery caused by war, he adopted Buddhist principles and renounced violence. Ashoka became a supporter of Buddhism, encouraging its dissemination both within and outside his empire.

12.6.3 Dhamma Policy

The concept of Dhamma, encompassing moral and ethical principles guiding people's behavior and state administration, is closely associated with Ashoka's reign. He commanded his subjects to practice compassion, tolerance, and respect for all living beings, and he promulgated decrees endorsing the Dhamma.

12.6.4 Rock and Pillar Edicts

Ashoka spread his message of moral leadership and social welfare through his edicts, inscribed on pillars and rocks throughout the empire. These decrees addressed various issues, including religious tolerance, animal welfare, environmental preservation, and social justice. To better serve his subjects, Ashoka established several welfare facilities, such as veterinary clinics, dispensaries, and hospitals. He also advocated for building roads, rest areas, and irrigation projects to enhance infrastructure and promote trade and transportation.

12.6.5 Legacy

Even after his death, Ashoka's influence persisted. During his rule, ancient India underwent a profound period of political, religious, and cultural change. His acceptance of Buddhism and his global inspiration for peace and harmony continue to resonate.

All things considered, Ashoka is regarded as a kind and visionary leader whose reign embodied the values of moral leadership and social responsibility. His contributions to the globalization of Buddhism and the development of Indian civilization have had a profound impact on history.

12.6.6 Hierarchical Administration

To manage his enormous empire, King Ashoka of the Mauryan Empire, who reigned from roughly 268 to 232 BCE, created a highly developed administrative structure. King Ashoka occupied the highest position in this hierarchy, holding ultimate power over the entire empire. Below him were viceroys or provincial governors in charge of specific areas of the empire. These governors selected administrators to oversee districts within their provinces. The king represented the highest level of authority, with each subsequent level reflecting the overall structure, creating a hierarchical system that resembled a fractal pattern.

12.6.7 Decentralized Governance

Despite the king's centralized authority, there was a significant amount of decentralization of power under Ashoka's rule. Rajas, who oversaw smaller provinces or cities within the empire, were responsible for local governance. These rajas had a degree of autonomy when it came to managing local affairs, such as taxation, law enforcement, and public welfare. This process of decentralization mirrored a fractal pattern, where the centralized authority structure was reflected in smaller units of governance throughout the empire.

12.6.8 Propagation of Dharma

One of King Ashoka's greatest achievements was spreading dharma, or moral and ethical values, throughout his empire. Ashoka's policies of social justice, religious tolerance, and moral leadership were communicated via his rock and pillar edicts. The king served as the primary source of authority in this hierarchical system of dharma dissemination, with his decrees trickling down to local officials and communities. Consequently, the spread of dharma followed a self-similar pattern, where the king's values influenced governance at multiple levels in society.

Through the lens of self-similarity and hierarchical administration, we can analyze King Ashoka's reign and observe how historical governance structures develop fractal-like patterns. Although these conceptual parallels have no direct mathematical application, they provide insights into the structure and dynamics of historical empires such as the Mauryan Empire. Known by many as Ashoka the Great, he was an ancient Indian emperor who governed the Mauryan Empire from 268 to 232 BCE and is regarded as one of the most important kings in Indian history. Ashoka was the son of Emperor Bindusara and the grandson of Chandragupta Maurya, the founder of the Mauryan Empire.

12.6.9 Military Conquests

Early in his rule, Ashoka launched military expeditions to expand the Mauryan Empire. His successful conquests extended the empire's borders over a significant portion of the Indian subcontinent. However, after the bloody Kalinga War, which resulted in substantial loss of life, Ashoka underwent a dramatic change of heart. Moved by the suffering caused by war, he adopted Buddhist principles and renounced violence. Ashoka became a supporter of Buddhism, encouraging its dissemination both within and outside his empire.

12.6.10 Dhamma Policy

A concept known as Dhamma, which included moral and ethical rules guiding people's actions and state governance, is associated with Ashoka's reign. He proclaimed the Dhamma and urged his followers to treat all living beings with kindness, tolerance, and respect. Ashoka's decrees, inscribed on stones and pillars throughout the empire, promoted social welfare and moral leadership. These decrees addressed numerous issues, including social justice, environmental preservation, animal welfare, and religious tolerance.

12.6.11 Welfare Measures

For the benefit of his subjects, Ashoka established hospitals, dispensaries, and veterinary clinics, among other welfare initiatives. He also supported the construction of roads, rest areas, and irrigation projects to enhance infrastructure and promote trade and transportation

12.6.12 Legacy

Long after his death, Ashoka's influence continued to be felt. During his rule, ancient India experienced a profound period of political, religious, and cultural change. His global inspiration for peace and harmony, along with his acceptance of Buddhism, remains relevant today. Ultimately, Ashoka is remembered as a compassionate and visionary ruler whose reign embodied the principles of social responsibility and moral leadership. His contributions to Indian civilization and the globalization of Buddhism have had a significant impact on history. While historical records do not provide a comprehensive list of Ashoka's ministers or cabinet members, it is known that he, like other historical Indian emperors, had a council of ministers and trusted advisors who supported him in governance and decision-making.

These ministers likely included members of noble families, reliable officials, and advisors with diverse expertise in areas such as finance, administration, diplomacy, law, and military affairs. Although the precise composition of Ashoka's council remains unknown, historical records illuminate the overall framework of government during his reign. The important individuals who might have served in Ashoka's administration are as follows:

1. **Prime Minister or Chief Advisor:** This individual would have been Ashoka's highest-ranking council member, responsible for managing the government and advising the king on important issues.
2. **Ministers of Finance:** He appointed officials tasked with overseeing the empire's financial operations, including economic policies, revenue collection, and taxation.
3. **Ministers of Foreign Affairs:** Advisors responsible for maintaining diplomatic relations with foreign nations and neighboring states.
4. **Military Commanders:** Generals and military commanders in charge of managing the army and leading military operations under Ashoka's direction.
5. **Judicial Officials:** Individuals responsible for enforcing the law and resolving disputes between parties within the empire.
6. **Religious Advisors:** Given Ashoka's endorsement of religious tolerance and his embrace of Buddhism, he likely had advisors knowledgeable in religious matters who helped formulate policies related to religious affairs.

Even though the precise identities of Asoka's ministers are unknown, it is clear that he relied on a group of knowledgeable and trustworthy advisors to govern his vast empire. These officials were essential for implementing Ashoka's policies, maintaining order, and ensuring the welfare of his subjects. His army likely included many capable generals and military commanders who played key roles in the empire's military campaigns and conquests, with Prince Tissa being one notable figure. However, specific information about individual military commanders in Ashoka's army may be scarce due to the limitations of historical records.

During his reign, Asoka likely had several religious advisors who were pivotal members of his cabinet, especially after he converted to Buddhism and embraced the

Dhamma (the moral code). While the identities of these advisors may not have been thoroughly recorded, historical accounts suggest that Ashoka sought guidance from Buddhist scholars and monks on both religious and political matters. The Buddhist monk Upagupta was one of Ashoka's most well-known religious advisors, held in high regard within the Buddhist community and believed to have played a crucial role in Ashoka's conversion to Buddhism. According to Buddhist customs, Upagupta helped guide Ashoka toward the Dhamma path, encouraging him to adopt non-violence, compassion, and religious tolerance. The king's rock and pillar edicts, which frequently reference Buddhist concepts and teachings, exemplify how Upagupta influenced Ashoka's religious beliefs and practices. Ashoka's commitment to advancing the material and spiritual well-being of his subjects, as directed by his Buddhist advisors, is reflected in these inscriptions.

Ashoka's noteworthy accomplishments in architecture are listed as follows:

1. **Ashoka's Pillars:** Throughout his empire, Ashoka constructed numerous pillars, many of which bore inscriptions outlining his moral precepts, guiding policies, and other details. These polished granite or sandstone pillars stand out for their remarkable height, fine workmanship, and symbolic significance. The pillars were deliberately positioned in key political and cultural areas to serve as symbols of Ashoka's dominance.
2. **Sarnath Pillar:** Located near Varanasi in present-day Uttar Pradesh, India, the Sarnath Pillar is one of the most well-known pillars built by Ashoka. Renowned for its capital, the Sarnath Pillar represents Ashoka's imperial rule and devotion to the Dhamma, featuring four lions standing back to back. The lion capital of the Sarnath Pillar is currently the national emblem of India.
3. **Stupas:** Throughout his empire, Ashoka ordered the construction of numerous stupas, or Buddhist monuments. These structures were built to honor significant moments in Buddhist history or to house Buddha artifacts. The Great Stupa at Sanchi is the most well-known of these stupas; it was initially commissioned by Ashoka and later enlarged and embellished by succeeding kings. The Great Stupa at Sanchi is considered one of the finest examples of Buddhist art and ancient Indian architecture
4. **Rock Edicts:** Ashoka left behind more than just architectural masterpieces when he carved inscriptions onto naturally occurring rock formations throughout his empire. These rock edicts communicated Ashoka's policies, moral precepts, and directives to his subjects. The rock edicts served as lasting testaments to Ashoka's commitment to responsible governance and religious acceptance.
5. **Roads and Infrastructure:** While not strictly architectural, Ashoka's reign saw significant investments in rest houses, roads, and bridges (referred to as "Ashokan edicts"). These infrastructure projects enhanced the economic prosperity and administrative efficiency of the empire, facilitating trade, travel, and communication within its borders.

All things considered, Ashoka's architectural accomplishments are a testament to his support of Buddhism, his dedication to moral leadership, and his desire to leave a lasting legacy of harmony and peace. His architectural creations continue to be researched and appreciated for their creative, historical, and cultural significance in the development of ancient Indian civilization. There is no evidence that the fractal patterns on Ashoka's pillars were intentionally incorporated into their design or construction. The majority of the decorative and symbolic elements on the pillars erected during Ashoka's reign reflect ancient Indian culture and religion. In Ashoka's pillar designs, the capital, which frequently features elaborate carvings and inscriptions, sits atop a cylindrical shaft. The most well-known example is the Sarnath Pillar, which showcases four lions standing back to back in the capital, supporting a platform that holds a Dharmachakra, or wheel of law. Although the capital's design includes fractal-like elements such as symmetry and repetition, it is unlikely that the architects of ancient India were aware of or intentionally employed fractal geometry in their creations.

The mathematical concept of fractal geometry was not developed until much later, and there is no historical evidence indicating that the ancient builders of India were familiar with fractals. Instead, it is more likely that the architectural elements and decorative motifs on Ashoka's pillars were influenced by royal insignia, religious symbolism, and popular artistic conventions of the era.

12.7 THANJAVUR TEMPLE

Ancient India's architectural accomplishments are impressive in their own right, but any similarities to fractal patterns are likely accidental rather than the result of intentional fractal design principles. The Brahadeeswarar Temple is renowned for its majesty, technical achievements, and aesthetic beauty; however, there is no concrete evidence that Raja Raja Chola purposefully incorporated fractal-like self-similar structures into the temple's design. Its enormous size, intricate carvings, and innovative engineering solutions—such as constructing the temple's imposing vimana (shrine) from a single stone block—contribute to its status as an architectural masterpiece. Certain features of the Brahadeeswarar Temple may unintentionally exhibit some degree of self-similarity, like the repetition of specific motifs or patterns in its sculptures and decorations. Indian temple architecture often employs this recurring motif for decorative, symbolic, and religious reasons.

While the Brahadeeswarar Temple is undoubtedly a masterpiece of ancient Indian architecture, any perceived similarities to fractal self-similarity patterns in its designs are most likely coincidental rather than intentional. Nevertheless, scholars, historians, and tourists continue to admire the temple, which stands as a testament to the Chola dynasty's creativity and artistic brilliance. To gain in-depth knowledge about the architecture, sculptures, and design of the Brahadeeswarar Temple (Periya Kovil), you can explore various sources, including books, scholarly articles, online resources, and visits to the temple itself. Here are some actions you can take to learn everything there is to know about the temple.

12.7.1 BOOKS AND ACADEMIC PUBLICATIONS

Seek out academic publications and scholarly books about Indian temple architecture, particularly that of the Brahadeeswarar Temple and Chola architecture.

Notable writers on this subject include George Michell, Adam Hardy, and others. Look for relevant books at university libraries, bookstores, or online resources. Online Resources: Utilize credible websites dedicated to Indian art and architecture, as well as scholarly journals and research papers. Academic databases like JSTOR and Google Scholar, along with resources such as the Archaeological Survey of India (ASI), can provide valuable information. Archaeological Survey of India Documentation: The Brahadeeswarar Temple has been extensively researched and documented by the ASI. To gain comprehensive details about the temple's architecture, sculptures, and historical significance, peruse their publications, reports, and online archives. Guided Tours and Lectures: Take advantage of opportunities to join knowledgeable guides or scholars on guided tours of the Brahadeeswarar Temple. These tours often provide in-depth explanations of the architectural elements, decorative patterns, and symbolic significance of the temple. Seeing the Temple: If possible, visit the Brahadeeswarar Temple in person to fully appreciate its sculptures, architecture, and design elements. Take your time to stroll around the temple, observing its mandapa (hall), vimana (tower), sculptures, inscriptions, and ornamental themes.

12.7.2 Documentaries and Audiovisual Resources

Watch documentaries or audiovisual presentations about the Brahadeeswarar Temple to enhance your understanding. Many documentaries feature expert interviews, visual tours of the temple, and analyses of its architectural and historical significance.

12.7.3 Expert Consultants

Reach out to scholars, historians, or experts specializing in Indian art and architecture for additional insights and guidance. They may recommend specific resources or offer valuable interpretations of the temple's architectural features and sculptural motifs. Combining these approaches will deepen your understanding of the Brahadeeswarar Temple's architecture, sculptures, and design, as well as its artistic, cultural, and historical significance in relation to the Chola dynasty and Indian temple architecture.

12.8 FRACTAL SCULPTURES IN CHOLA DYNASTY

This book examines the origins and development of Indian temple architecture, including sculpture and carvings. It covers temples built by various ancient and medieval dynasties throughout Indian history, particularly the Pallava, Pandya, Chola, Hoyasala, and Nayaka. The work features over a hundred plates depicting temples and carvings, including images of pillars and cave facades. The illustrations also include maps, plans of caves, and representations of viharas and caityas. The study encompasses temples located in various states of India. To explore the evolution of temple architecture over time, the research addresses variations in plans and elevations, as well as innovative construction techniques.

Figure 12.1 depicts the sculpture of the Brahadeeswarar Temple. Figure 12.2 depicts the outer surface area of the Brahadeeswarar Temple.

FIGURE 12.1 Sculpture of the Brahadeeswarar Temple.

FIGURE 12.2 Outer sculptures of the Brahadeeswarar Temple.

12.9 THE STUCCO PICTURE ON THE VIMANA OF THANJAVUR'S BRAHADEESWARAR TEMPLE

A stucco sculpture of a Chinese-looking trader, emissary, or ruler was created in the Brahadeeswarar Temple in Thanjavur, likely to commemorate the friendly emissary exchange and relationships between Raja Raja and the Chinese court. Installing this figure would have required Raja Raja's approval, as the temple was his personal initiative. This stucco figure is placed inside the kudu arch of the makara-torana ornamentation of a panjarakoshta, located on the second tala of the northeast corner of the vimana. The elaborate makara-torana embellishments adorn the panjarakoshta. At the base of the kudu arch is a bust of a man wearing a cap or hat. Standing on either side of the kudu arch are open-mouthed makaras, from which warriors wielding swords and shields emerge. The illustration depicts how the vayaliface at the top merges with the floriated tails of the makaras as they rise higher.

Figure 12.3 is the sculpture of the Chinese-looking trader.

The region around the floriated border above the kudu arch is decorated with horizontal frescoes of Bhuta, Simha, and Pushpa Hara, clearly visible from the front courtyard of the Chandikeshwara Temple. The Chinese man in the hat is depicted with his hands folded and resting on a pot. He has a broad, mongoloid face with a relatively sharp nose characteristic of Chola style. His teeth sparkle in a wide smile, complemented by a moustache and a goatee. The moustache, growing timidly beneath the nasal septum or philtrum, becomes more luxuriant as it extends outward, exhibiting typical mongoloid traits. He is wearing a half-sleeved "Chinese" collar shirt and a hat or cap with small flaps on the sides of his head. The image of the Chinese man with the hat (Pl. 2) shows him reclining on a support with his hands folded.

FIGURE 12.3 Chinese-looking trader.

He has a broad, Mongoloid face. His nose is relatively sharp for a Mongoloid face because of the Chola style. His teeth sparkle in his wide smile. He has a mustache and a goatee. Beneath the nasal septum, or philtrum, the mustache grows timidly; however, as it extends out to the sides, it becomes more luxuriant and characteristic of Mongoloid traits. His headwear consists of a cap or hat with short side flaps, and his shirt features a half-sleeve "Chinese" collar.

There are stone sculptures on the 216-foot-tall vimana of the Brahadeeswarar Temple that seem to represent foreign figures, including what some have described as Chinese men wearing characteristic conical caps. The outside walls of the temple are covered in a wide variety of carved images, which include these figurines. The existence of these figures has prompted a number of theories and interpretations. According to some academics, these sculptures depict diplomats or foreign businessmen from distant places, representing the Chola dynasty's links and trading relations with diverse regions of Asia, including Southeast Asia and China. The sculptures may illustrate the Chola Empire's global reach at its height. It's crucial to remember that these interpretations are predicated on historical context and visual evidence, and specialists may have different opinions about the precise identity and significance of these sculptures. The artwork of the Brahadeeswarar Temple displays a fusion of indigenous and foreign artistic elements and is rich in symbolism and cultural influences from the era. It has been designed with characteristics of self-similarity.

12.10 CONCLUSION

In architecture, fractals offer an enthralling fusion of science and art, providing designers with a flexible toolkit to create aesthetically magnificent and operationally effective structures. By utilizing fractal geometry, architects can produce complex patterns, organic forms, and efficient use of space. A major benefit of using fractals in architectural design is their ability to scale. Architects can maintain harmony and coherence at multiple levels of detail, from the overall building arrangement to the smallest ornamentation, thanks to the self-similarity of fractal patterns at various sizes. This characteristic enhances the visual appeal of the architectural composition while imparting a sense of coherence and unity.

By emulating tree branching systems, intricate snowflake forms, or irregular coastal outlines, architects can evoke feelings of balance, a sense of connectedness to the natural world, and awareness of biophilic design principles. Fractals not only possess aesthetic appeal but can also enhance the practical aspects of building. For instance, fractal-based patterns can improve structural stability, encourage natural lighting and ventilation, and maximize space efficiency. By applying fractal principles, architects can design structures that are not only visually striking but also robust, sustainable, and conducive to human well-being. Overall, the incorporation of fractals in architecture represents a captivating blend of biology, mathematics, and the arts. As the potential of fractal geometry continues to be explored, we can anticipate seeing increasingly inventive and breath-taking architectural works that maximize functionality, harmonize with the natural environment, and foster appreciation for the inherent beauty of mathematical patterns.

REFERENCES

1. Abboushi, B., Elzeyadi, I., Taylor, R., and Sereno, M., Fractals in architecture: The visual interest, preference, and mood response to projected fractal light patterns in interior spaces, *Journal of Environmental Psychology*, Vol. 61, pp. 57–70 (2019).
2. Fleron, J. F., A note on the history of the Cantor set and Cantor function, *Mathematics Magazine*, Vol. 67(2), pp. 136–140 (1994).
3. Joye, Y., Fractal architecture could be good for you, *Nexus Network Journal*, Vol. 9, pp. 311–320 (2007).
4. Kitchley, J. L., Fractals in architecture, *Architecture and Design-New Delhi*, Vol. 20(3), pp. 42–51 (2003).
5. Lorenz, W. E., Andres, J., and Franck, G., Fractal aesthetics in architecture, *Applied Mathematics & Information Sciences*, Vol. 11(4), pp. 971–981 (2017).

13 Fractal Neural Networks and the Industrial Use Cases

13.1 INTRODUCING NEURAL NETWORKS

A neural network (NN), a machine learning (ML) model, is a fascinating creation that mimics the function and structure of the human brain. Thanks to its ability to learn from training data, it can make decisions similar to those made by the human brain. This learning process mirrors how biological neurons collaborate to predict, identify patterns, and make decisions. The deeper we explore the world of NNs, the more we recognize their potential and the endless possibilities they offer.

NNs automatically extract decision-enabling features from data. Generally, every NN consists of multiple layers, including an input layer and an output layer, with one or more hidden layers in between. Each layer is made up of nodes or neurons, and each node connects to other nodes in the subsequent layer. Each node has an associated weight and threshold. If the output of any individual node exceeds the specified threshold, that node is activated and can send data to the next layer of the network. If not, no data are passed to the next layer.

NN's rely on training data to learn and improve their accuracy over time. While humans may take hours to complete these data-intensive and error-prone tasks, NNs can finish them in minutes. One prominent example of an NN model is Google's search engine. To distinguish artificial NNs (ANNs) from biological NNs, we refer to the former as ANNs. When numerous hidden layers are involved in solving complex problems, ANNs are termed deep NNs (DNNs), which are essential deep learning (DL) models.

Euclidean classical geometry uses idealized abstractions to model real-life objects, but their structures have become more complex. Fractal objects cannot be accurately described without referencing infinity due to their inherent complexity. Euclidean geometry falls short in characterizing phenomena like crystal growth, chaotic motion, and turbulence owing to their high irregularity. Consequently, these phenomena became subjects of fractal geometry. In this chapter, we will delve into fractal dimension analysis (FDA) and fractal NNs (FNNs), as fractal analysis has provided solutions to many complex problems.

How Do NNs Work? It is logical to consider each node as a linear regression model ($y = ax + b$), where x is the input value, a is the weight, b is the bias, and y is the output. The output is one if ax1 + b is greater than zero; it is 0 if $ax_1 + b$ is less than zero. There can be many input values, each with its corresponding output values. Afterward, the output is passed through an activation function, which determines the

DOI: 10.1201/9781003481096-13

final output. If this output exceeds a given threshold, the node "fires" (or activates), passing data to the next layer in the network. This process of passing data from one layer to the next defines this NN as a feedforward network.

The input layer includes email content, sender information, and subject. These inputs, each multiplied by adjustable weights, are passed through hidden layers. Through training, the network learns to recognize patterns that indicate whether a particular email is genuine or fake. This real-world application of NNs—differentiating between legitimate and spam emails—is just one example of their potential. The iterative refinement of weights through backpropagation is a testament to their learning capabilities and adaptability.

13.2 FORWARD PROPAGATION

- **Input Layer:** Each node in a NN's input layer represents a feature. For instance, if there are ten features, there will be ten nodes in the input layer.
- **Weights and Connections:** The weight of each neuronal connection indicates its strength. These weights are adjusted during training.
- **Hidden Layers:** Each neuron in the hidden layers processes inputs by multiplying them by weights, summing the results, and then passing them through an activation function, which introduces nonlinearity. This setup enables the network to recognize intricate patterns.
- **Output:** The final output is produced by repeating the process until the output layer is reached.

13.3 BACKPROPAGATION

- **Loss Calculation:** The network's output is evaluated against the actual target values, and a loss function is used to compute the difference. For regression problems, the mean squared error is commonly used as the cost function.
- **Gradient Descent:** The network employs gradient descent (the derivative of the loss with respect to each weight) to minimize the loss. The weights are adjusted accordingly. Through backpropagation, the weights at each connection are refined.
- **Training:** Training involves using different data samples and comprises forward propagation, loss calculation, and backpropagation. When performed iteratively, training enables the network to learn and discover hidden yet valuable patterns in the data.
- **Activation Functions:** These functions introduce nonlinearity.

Artificial intelligence (AI) services, solutions, and systems are implemented through ML and DL algorithms. These algorithms parse input data, learn from it, and then apply their knowledge to make intelligent decisions. There are supervised, unsupervised, and semi-supervised learning algorithms. Reinforcement learning is another promising learning method.

13.3.1 Learning with Supervised Learning

In supervised learning, the NN learns by accessing both input–output pairs. The network generates outputs based on inputs without considering the surrounding context. An error value is calculated by comparing the generated outputs to the actual outputs. By appropriately adjusting the connection weights, model performance can be improved to an acceptable level.

13.3.2 Learning with Unsupervised Learning

Clustering and association tasks are effectively addressed through unsupervised learning.

13.3.3 Demystifying DNNs

DNNs are an advanced version of NNs and represent a powerful implementation technology for next-generation AI systems. DNNs can learn the complexities and insights hidden within data collections. They can automatically extract decision-enabling features without the need for explicit programming to instruct them on how to learn from the input data. With the emergence of AI-centric processing units, DNNs are being explored and tested for various industrial applications, achieving remarkable success rates. DNNs are inspired by the structure and function of biological neurons in the human brain and require substantial data to demonstrate human-like intelligence in their decisions, predictions, suggestions, generations, and conclusions.

DNNs possess the inherent capability to manage big data effectively. They eloquently represent complex problems and streamline problem resolution through well-designed architectures. The ability to incorporate multiple layers of interconnected nodes within DNNs is a game changer for solving complicated issues and visualizing next-generation use cases across personal, social, and professional domains. DNNs enable DL, which has achieved significant success in applications such as facial recognition, object detection, speech recognition, text generation, summarization, and translation. The field of cloud AI is thriving with the rapid maturity and stability of GPUs, TPUs, ASICs, FPGAs, NPUs, and other technologies.

Software engineering differs significantly from AI model engineering. NN models are distinct from traditional software applications, as NN models can continuously learn even with missing or inaccurate components. Experts argue that system architectures capable of learning complex tasks or detecting anomalies can be transformative. AI is versatile, leading multiple industrial sectors to actively explore and experiment with its disruptive potential. While traditional software engineering is typically formal, AI model engineering tends to be more informal.

One well-known application of DNNs is in self-driving cars. These vehicles gather various types of information about their road environment, including the locations of other vehicles and any obstacles. These data can be meticulously collected and processed through an appropriate DNN to make decisions. The knowledge extracted by the DNN helps self-driving vehicles navigate their surroundings effectively.

There are several types of NNs, each with distinct strengths and limitations. Some networks excel at solving problems with precise input–output mapping, while others are better suited for processing data with a grid-like structure, such as images.

13.4 DIFFERENT TYPES OF DNNS

The field of NNs is rapidly evolving, with many types emerging to address a variety of problems. In this section, we will highlight some of the most prominent NNs [1].

1. **Feedforward NNs, or Multi-Layer Perceptrons (MLPs),** consist of an input layer, one or more hidden layers, and an output layer. MLPs are composed of sigmoid neurons since most real-world problems are nonlinear. They are effective at solving problems with a clear relationship between input and output. For example, if presented with an image of a cat, a feedforward NN can quickly and accurately recognize it as a cat. However, feedforward NNs can also learn and recognize complex relationships and patterns.
2. **Convolutional NNs (CNNs)** are also feedforward networks, primarily used for computer vision (CV) tasks. CNNs excel at matrix multiplication, which is essential for pattern identification and recognition within images represented in matrix form. CNNs utilize convolutional layers (which can be entirely connected or pooled) to automatically learn hierarchical features from input images, enabling effective image recognition and classification. These convolutional layers generate feature maps that capture regions of the image, which are subsequently processed for nonlinear analysis.
3. **Recurrent NNs (RNNs)** are distinguished by their feedback loops. RNNs perform exceptionally well on sequential and time series data, making them suitable for tasks involving ordered information, such as words in a sentence. They excel at understanding intricate relationships between different pieces of data and are appropriate for applications with critical contextual dependencies, such as time series prediction, language translation, and speech recognition. In the RNN model, each node acts as a memory cell, facilitating the ongoing computation and execution of operations.

   This NN begins with the same front propagation as a feedforward network but then retains all processed information for future use. If the network's prediction is incorrect, the system self-learns and continues working towards a correct prediction during backpropagation.
4. **Long-Short-Term Memory (LSTM)** is a type of RNN designed to overcome the vanishing gradient problem in training RNNs. It employs memory cells and gates to selectively read, write, and erase information.
5. **Generative Adversarial Networks (GANs)** consist of two NNs (a generator and a discriminator) that collaborate to produce synthetic data that appears real [2]. GANs are particularly effective in creating realistic images, audio, video, art, music, etc.
6. **Autoencoder NNs** reduce data complexity, making it easier and faster to learn essential features. They are used in tasks such as image and speech recognition.

Importance of NNs: As noted above, the unique contributions of NNs lie in identifying patterns, solving intricate problems, and adapting to changing environments. Their self-learning capability is leveraged to address complex issues and envision next-generation industrial applications. AI depends on the extraordinary power of NNs to remain accurate and relevant to society.

13.5 ADVANTAGES OF NNs

NNs offer numerous business and technical benefits. Any data converted to numbers can be used with an NN, as NNs are mathematical models that utilize approximation functions. NNs can model nonlinear data (such as images) and handle multiple inputs concurrently. With more features, NNs become an excellent tool. They are reliable because they can decompose classification problems into a layered network of more specific elements. Once trained and optimized, NNs can make predictions quickly and adapt to any number of inputs and layers.

- **Adaptability:** In NNs, the relationship between inputs and outputs is often complex or not well-defined. Therefore, NNs must adapt to new situations and learn from data, making them suitable for both regression and classification problems.
- **Pattern Recognition:** NNs excel in recognition tasks, such as image and speech recognition.
- **Parallel Processing:** NNs can perform parallel processing, allowing them to execute multiple tasks simultaneously.
- **Nonlinearity:** NNs can model and understand complex relationships in data by leveraging nonlinear activation functions.
- **Fault Tolerance:** NNs possess fault tolerance, meaning that the failure of one or more neurons will not halt output generation.
- **Gradual Corruption:** The performance of an NN degrades gradually over time, rather than abruptly when a problem occurs.
- **Unrestricted Input Variables:** Input variables are not limited in any way regarding their distribution.
- **Unorganized Data Processing:** NNs can process, sort, and categorize large datasets efficiently.
- **Ability to Generalize Data:** NNs can infer unseen relationships from data they have not encountered before.

13.6 DISADVANTAGES OF NNs

NNs have some inherent limitations:

- **Computationally Intensive:** Large language models (LLMs) with billions of parameters/weights are trained on trillions of tokens. LLMs serve as a prominent example of DNNs. The pre-training and fine-tuning of LLMs consume substantial computational resources and energy.

- **Black Box:** NNs are typically considered black box models. Consequently, the transparency and trustworthiness of the decisions produced by NNs are often questioned.
- **Overfitting:** While NNs are designed to learn from data to identify useful patterns, they can sometimes memorize the training material, leading to overfitting. This results in a failure to generalize to unseen data.

NNs make decisions based on a set of qualities (data and features), values, or requirements at a given moment. Occasionally, NNs may arrive at and articulate incorrect decisions. Again, NNs are typically regarded as black box models, making it difficult to understand how independent input variables influence dependent output variables. As is widely known, the computational requirements for training NNs are significant, and the demand for AI-centric processors is increasing considerably. Finally, DL algorithms can be prone to overfitting, meaning they may perform well on training data but poorly on unseen data. Proper data splitting and regularization techniques can help mitigate this issue.

13.7 THE FUTURE OF NNS

Some potential future developments in NN technologies include the integration of the following:

- Fuzzy logic, which allows for more than just true or false values, and can be designed for various applications.
- Pulsed neural, which use the timing of pulses to transmit information and perform computations
- Neurosynaptic architectures, which function more like a biological brain than a traditional computer.

NNs show promise in empowering robotic applications to make intelligent decisions. They may also provide opportunities for human–machine brain melding, where NNs have the potential to connect human brains with AI.

13.8 ML VS. DL

A blog by an expert states that DL is a subset of ML. DL is more advanced and nuanced than traditional ML, allowing for the tackling of complex problems through DL algorithms. It achieves remarkable power and flexibility by inherently learning to represent the world as a nested hierarchy of concepts. A DL algorithm can learn categories incrementally through its hidden layer architecture, starting with low-level categories like letters, followed by higher-level categories like words, and then more complex categories like sentences. In the case of image recognition, the process begins by identifying light and dark areas, then categorizing lines, and finally recognizing shapes to complete face recognition. Each neuron or node in the network represents one aspect of the image, and together they form a complete representation. Each node is assigned a weight that reflects the strength of its relationship to the output, which changes as the model develops.

Domain experts must identify and finalize decision-enabling features to solve business problems using ML algorithms. However, DL algorithms inherently have the power to learn high-level features from data incrementally, automating the feature extraction process. DL methods facilitate end-to-end problem resolution.

To leverage ML techniques effectively, explicit descriptions must be formulated initially. These descriptions break down the problem into different components to be solved individually, with their results combined in the final stage. For instance, in a multiple object detection problem, DL frameworks like take the image as input and output the location and names of identified objects. Conversely, when using an ML algorithm like the support vector machine, the resolution process begins with a bounding box detection algorithm that identifies all objects using the histogram of oriented gradients. This output becomes the input for the learning algorithm to recognize relevant objects.

Generally, a DL algorithm takes a long time to train due to the many parameters involved. In contrast, ML algorithms typically take only a few seconds to train. However, during the testing phase, DL algorithms often require significantly less time for testing. In some cases, the testing time for certain ML algorithms increases as the data size grows.

Regarding explainability and interpretability, ML algorithms (such as decision trees and logistic regression) perform well. However, DL algorithms often result in black box models, leading to questions about transparency. Humans cannot easily interpret how DL models arrive at their decisions, nor can these algorithms explain the reasoning behind their recommendations. Recently, an additional abstract layer of explainability has been incorporated into DL models to clarify their conclusions and predictions.

In summary, NNs consist of multiple layers of interconnected nodes, referred to as artificial neurons. These neurons, like the biological neurons in our brains, can receive, process, and transmit data. In the future, we can expect a greater variety of application-specific NNs. Advanced NNs will be developed for deployment in edge devices such as robotics, drones, and brain–machine interfaces. Currently, we primarily utilize artificial narrow intelligence, but NNs will undergo significant advancements to fulfill the long-term goal of artificial general intelligence or superintelligence. In other words, while we are experiencing generative AI today, we are progressing toward creative AI.

Furthermore, high-end platform solutions are making significant strides in AI model engineering, evaluation, optimization, deployment, observability, and improvement. The widespread establishment and maintenance of cloud-native infrastructure modules (including compute, storage, network, and security) globally facilitate the resolution of DL challenges.

13.9 DEMYSTIFYING GRAPH NNs (GNNs)

Graphs are non-Euclidean data structures increasingly used to simulate complex real-world scenarios, such as brain networks, traffic networks, biological networks, citation networks, and social networks. A set of objects and the connections or relationships between them are neatly expressed as a graph. Graphs are gaining special significance as they are uniquely efficient in representing and exchanging multi-structured data. They help retrieve semantically rich and syntactically correct information from graph-structured data, making them invaluable for tackling various complicated problems across industry verticals. Numerous digital transformation initiatives of businesses and national governments worldwide are being simplified and accelerated through the advancements unveiled in graph technology.

DL algorithms have undoubtedly enhanced applications in computer vision and speech recognition. These algorithms can extract high-level features from data by passing it through multiple nonlinear layers, efficiently handling Euclidean-structured data such as tabular data, images, text, and audio. However, classical DL algorithms are often less suited for processing graph data. As a result, technologists have explored ways to integrate graph data into DL algorithms to extract actionable insights. Experts advocate for strategically combining graph-structured data and DNNs. This convergence is being termed graph NNs (GNNs), hailed as a game-changing phenomenon for addressing a variety of next-generation business requirements and creating sophisticated applications. In essence, robust NNs are being developed and employed on graph-represented data to accomplish larger and more complex tasks.

Researchers have built purpose-specific GNNs to tackle well-known problems such as traffic prediction, drug discovery, fake news detection, and supply chain optimization.

Graph data are complex for existing ML algorithms. Conventional ML and DL tools are specialized for simple data types. For example, images with a consistent structure and size (considered fixed-size grid graphs) can be easily processed by CNNs. Text and speech are sequential and, therefore, considered as line graphs. However, there are complex graphs without a fixed form. Unordered nodes with varying sizes and numbers of neighbors are prevalent in graph data. In traditional ML algorithms, it is assumed that instances are independent of one another. In contrast, graphs exhibit a different scenario, where each node is connected to other nodes. This relationship is indicated through links.

GNNs are NNs designed to process graph data, providing an efficient and elegant mechanism for performing node-level, edge-level, and graph-level prediction tasks. The unique combination of graphical representation of data with the predictive power of machine and DL models has proven to be a game changer for the evolving era of knowledge.

Consider the example of classifying products sold in a store. Typically, we gather various product details (such as manufacturer name, location, date, price, customer feedback, etc.) to train an ML model. Each product can be represented as a node, with links established between products that are frequently purchased together. Other possibilities for linking different products also exist. Thus, a graph-based ML model provides more intuitive and nuanced product information. Due to their powerful expressive capabilities, graphs are receiving significant attention in ML. Each node is paired with an embedding, which establishes the node's position in the data space.

GNNs are topologies of NNs that operate on graphs. The primary goal of a GNN architecture is to learn an embedding that contains information about its neighborhood. This embedding can then be utilized to tackle various issues, including node labeling and node and edge prediction, among others. GNNs are a subclass of DL techniques specifically built to make inferences on graph-represented data. They can perform prediction tasks at the node, edge, and graph levels. The human brain's reasoning process is analogous to creating graphs based on daily experiences in the real world.

GNNs are strongly influenced by CNNs and graph embedding techniques. GNNs are utilized for node- and edge-level predictions as well as for various graph-based tasks. Compared to CNNs, GNNs can achieve optimal results due to their ability to handle arbitrary graph sizes and complex structures. The input graph is processed through a series of NNs, converting its structure into graph embeddings. This transition helps retain information about nodes, edges, and global context.

13.10 DEMYSTIFYING FNNs

Fractals are fascinating mathematical constructs characterized by self-similarity across scales. With a greater understanding of these structures, researchers are meticulously applying fractals across various fields, from natural phenomena to art. Fractals are geometric objects that exhibit self-similarity, meaning parts of the object resemble the whole at different scales. They possess intricate and irregular structures that defy traditional Euclidean geometry. Fractal patterns are widely found in nature, manifesting in the branching of trees and blood vessels as well as the contours of coastlines. Artists embrace fractals for their aesthetic appeal, incorporating them into paintings and sculptures. Below are the essential characteristics of fractals as described by experts

1. **Self-Similarity:** Fractals contain smaller copies of themselves at various scales, a hallmark of fractal geometry.
2. **Irregularity:** Fractals are irregular and may exhibit complex and chaotic shapes. This complexity arises from the repetition of patterns at smaller scales.
3. **Non-Integer Dimension:** Fractals are typically nonlinear. Unlike classical geometric objects with integer dimensions (e.g., lines, squares, and cubes), fractals have non-integer dimensions, known as the fractal dimension.
4. **Infinite Detail:** By zooming in on a fractal, we can discover increasingly intricate structures.

FNNs are becoming a key subject of study and research in many AI domains. With their complex repeating architectures at different scales, fractal networks are expected to significantly enhance the capabilities of NNs, leading to transformative use cases.

A fractal is defined by a pattern that repeats itself across scales. It consists of elements that are made up of the same fractal core architecture. A fractal network is an architecture constructed based on fractal expansion principles [3]. A functional block in an NN may exhibit a fractal architecture, characterized by fractal repetition at both micro and macro scales. Thus, there is an underlying symmetry in the design.

DNN architectures mimic the complex networks of neurons in the brain [4]. DNNs typically comprise billions of artificial neurons that exhibit behaviors similar to biological neurons. The connections between these neurons constantly change as they learn and unlearn. A DNN with fractal architecture will have multiple sub-levels, each with varying degrees of interconnections among the neurons.

Why Use FNNs? The uniqueness of any fractal network lies in its single expansion rule, which leads to an ultra-deep network. The network features subpaths of varying lengths, where each node interacts with others. With the addition of appropriate non-linearities and filters, continuous transformations can be enabled between layers.

With this new perspective, fractal networks can achieve higher complexity and exceed performance levels while maintaining low error rates [5]. They also possess certain intrinsic design advantages. An efficiency-optimized network will maintain efficiency throughout its sub-levels.

13.11 INDUSTRY USE CASES OF FNNs

- **Computationally Intensive Problems:** It is widely accepted that the greater the complexity of an NN, the more complex problems it can handle and solve. As the depth of an FNN, which is self-similar, increases, its computational power also increases manifold. Thus, ultra-deep FNNs have numerous advantages in solving complex problems compared to DNNs.
- **Advancing DL Use Cases:** The advancement of DNNs has led to several use cases (image recognition, object detection, speech recognition, language translation, etc.). These use cases can be substantially improved by incorporating fractal neural architecture.
- **FNNs for Fractal Data:** Conventional machine and DL algorithms can be used to learn from fractal data. However, experts point out those fractal algorithms can learn better from such data. That is, FNNs applied to fractal data can be more powerful and ground-breaking.
- **Natural Language Processing:** Natural language processing, understanding, and the generation of use cases, frameworks, and models are quickly evolving. In addition to information processing technologies and tools, we are entering the era of language processing platforms. With the maturity and stability of FNNs, the domain of natural language processing will be distinctly advanced. FNNs are promising and paramount in handling complex data and problems and bringing efficient solutions.

FNNs can guarantee big computational power with each layer in DNNs. FNNs mimic natural systems and use DNNs to achieve the power and intelligence of natural biological systems. DNNs can keep track of billions of parameters to perform complex tasks. This advancement accomplishes complicated and sophisticated tasks such as face recognition and language generation.

FNNs can ensure significant computational power with each layer in DNNs. FNNs mimic natural systems, leveraging DNNs to achieve the capabilities and intelligence of biological systems. DNNs can manage billions of parameters to perform complex tasks, accomplishing intricate and sophisticated operations such as face recognition and language generation.

13.12 FRACTAL GENERATION APPROACHES

Fractals can be generated using recursive algorithms or iterated function systems (IFSs).

Illustrating FDA

FDA is a robust metric for quantifying the complexity and self-similarity of fractal objects. The fractal dimension measures how a fractal fills space as it is iteratively divided into smaller copies of itself. Unlike classical dimensions (e.g., one-dimensional lines and two-dimensional squares), the fractal dimension can be non-integer, indicating the degree of space-filling irregularity. Here are a couple of methods for calculating fractal dimension:

Box-Counting Dimension (Minkowski–Bouligand Dimension): This method involves dividing the fractal into smaller boxes and counting the number of boxes needed to cover the fractal at different scales. Hausdorff Dimension: This method is used for more irregular fractals and measures how the "size" of an object changes as we zoom in or out.

13.13 APPLICATIONS OF FDA

This advanced analysis presents fresh use cases across multiple domains:

- **Geology:** Characterizing the roughness of geological formations and the irregularity of coastlines.
- **Biology:** Quantifying the complexity of biological structures such as lung airways and blood vessels.
- **Image Processing:** Enhancing image compression, texture analysis, and edge detection
- **Finance:** Analyzing self-similarity and volatility clustering in financial time series data.

Now, we will discuss how fractal dimension is applied to quantify computational complexities.

13.14 FRACTAL DIMENSION TO MEASURE COMPUTATIONAL COMPLEXITY

FDA, widely associated with the intricate geometries of fractal objects, is being skillfully used to characterize computational complexities. Generally, computational complexity focuses on time and space requirements, but the fractal dimension introduces a new perspective. FDA aids in thoroughly examining the structural properties of algorithms and problem instances. This analysis could offer a deeper understanding of the underlying patterns and irregularities in computational processes. FDA can elucidate the relationship between algorithms and their underlying implementations.

13.15 FDA FOR ANALYZING ALGORITHMIC STRUCTURES

This approach provides deeper insights into the capabilities of algorithms.

- **Algorithmic Structures:** Algorithms are often analogous to fractals. They possess structures that can exhibit self-similarity and complexity at multiple scales. By analyzing the fractal dimension of algorithmic structures, it is possible to gain insights into the inherent intricacies of algorithm design.
- **Algorithm Efficiency:** The fractal dimension is also used to accurately assess the efficiency of algorithms in solving specific problem instances.

Experts point out that algorithms with lower fractal dimensions may indicate higher self-similarity and efficiency in handling problem variations.

13.16 FDA IN PROBLEM INSTANCES

Applying FDA to problem instances brings noteworthy benefits.

- **Instance Complexity:** Each problem instance can be viewed as a fractal object with its own fractal dimension. Analyzing these dimensions provides clues about the complexity of any problem instance. This complexity may indicate how challenging it is for an algorithm to solve the problem instance.
- **Scaling Behavior:** The fractal dimension of problem instances can reveal how their complexity scales with varying input sizes. This understanding aids in predicting the computational resources required for solving instances of different sizes.

By incorporating FDA into computational complexity assessments, researchers and architects gain an additional criterion that simplifies algorithm selection. Algorithms that align with the fractal dimension characteristics of problem instances may exhibit superior performance and efficiency. Furthermore, FDA enables the dynamic adaptation of algorithms based on the complexity of problem instances. This adaptability enhances the efficiency of algorithms in real-world applications.

The FDA's application helps users make informed decisions about algorithm selection, optimize algorithm performance, and dynamically adapt to the complexities of real-world computational challenges.

13.17 MEASURING FRACTAL DIMENSION IN PROBLEM INSTANCES

Quantifying the fractal dimension of problem data simplifies a holistic understanding of algorithmic behavior. However, to measure the fractal dimension in problem instances, the data must be prepared appropriately.

- **Data Representation:** The best practice is to convert problem instances into appropriate data representations such as graphs, images, or numerical datasets.
- **Scaling:** It is important to ensure that data are scaled appropriately to represent different input sizes. For example, when analyzing graphs, it is crucial to scale the size and structure of the graphs accordingly.

Measuring the fractal dimension of problem instances can inform algorithm selection and optimization. It represents a practical approach to gaining insights into their computational complexities. By quantifying self-similarity and irregularity, we can achieve a more nuanced understanding of problem data and its implications for algorithmic behavior. This approach could enhance algorithm selection, optimization, and adaptability in the face of real-world computational challenges.

Implications for the Nature of Complexity: Applying FDA to computational complexity offers profound insights into the nature of complexity itself. This analysis reframes our perception of complexity.

1. **Beyond Linearity:** Traditional complexity measures often assume linear relationships between problem size and resource requirements. In contrast, FDA acknowledges the nonlinear, self-similar, and irregular nature of complexity.
2. **Hierarchy of Complexity:** The fractal dimension provides a metric for categorizing problems based on their self-similarity and irregularity, creating a hierarchy of complexity. This hierarchy extends beyond conventional complexity classes such as P and NP.

13.18 HIERARCHICAL COMPLEXITY

- **Hierarchical Structures:** FDA reveals hierarchical structures within computational problems. These problems can be grouped based on their fractal dimensions, highlighting the relationships and similarities between problem families.
- **Thresholds of Complexity:** FDA identifies complexity thresholds that signify critical points where problems transition from manageable to intractable. This knowledge enhances our understanding of the limits of efficient computation.

13.19 ALGORITHMIC ADAPTABILITY

FDA has significant implications for algorithm design and adaptability.

1. **Algorithm Selection:** Algorithms that adapt to the fractal dimension characteristics of problem instances may exhibit superior performance. This aligns algorithm selection with the inherent complexity of the problem at hand.
2. **Dynamic Algorithms:** The concept of fractal dimension encourages the development of dynamic algorithms that adjust their strategies based on the measured fractal dimension. Such adaptability enhances algorithm efficiency in real-world scenarios.

FDA extends beyond theoretical problems and can be applied to real-world data. Understanding the fractal dimension of data can aid in predicting computational requirements and optimizing data processing. This analysis spans various domains, including biology, geology, finance, and beyond. Insights from computational complexity can inform other fields grappling with complex and irregular data.

FDA aids in reconsidering the nature of complexity. Complexity is not solely a product of problem size but is deeply intertwined with the structure and self-similarity within problems. Recognizing this inherent complexity helps formulate more nuanced problem-solving strategies.

1. FDA as a straightforward computational approach to improve breast cancer histopathological diagnosis
2. FDA of resting state functional networks in schizophrenia from EEG signals
3. FDA for automatic morphological galaxy classification
4. Diagnosis system for hepatocellular carcinoma based on fractal dimension of morphometric elements integrated in an ANN
5. An algorithm for crack detection, segmentation, and fractal dimension estimation in low-light environments by NNs
6. FDA: A new tool for analyzing colony-forming units

IFSs are primarily used to construct fractals. The result of this construction is always self-similar. A fractal is a figure formed by the union of several copies of itself, where each copy is obtained through a function as part of a system of recursive functions. The fractal shape generated by this system of functions consists of several overlapping small copies of itself. This process is repeated ad infinitum. Each fractal consists of copies of itself, with self-similarity being the most essential characteristic of fractal shapes. The inverse operation is called fractal image coding. The code of the IFS is constructed through this process to generate a fractal image. Identifying the function code is one of the most critical challenges in ANNs. (More can be found in the paper "Activation Functions Effect on Fractal Coding Using Neural Networks" by Rashad A. Al-Jawfi.)

13.20 FNNs IN NETWORK SECURITY

The goals are to accurately perceive network security situations, predict development trends, and effectively defend against network attacks. The LSTM network is utilized as the basis for the network security situation awareness and prediction model. Furthermore, it is optimized using the genetic algorithm (GA) to enhance its global search capability. An FNN is then constructed based on fractal theory, which is employed in network security situation awareness to mitigate exploding or vanishing gradient problems. By using the fractal difference function as the activation function, gradient variation can be balanced and stabilized. Additionally, this approach improves the feasibility and effectiveness of the NN structure for network security situation awareness and prediction. The FNN under study holds practical significance for assessing the current network security situation and predicting its evolving trends, providing a reference for protecting internet operations from network attacks. Readers can find more information in the paper "Application of Fractal Neural Network in Network Security Situation Awareness."

The FNN constructed from NNs and fractal theory can perceive network security situations from multiple levels and angles, enhancing the recognition capabilities of the NN. Moreover, the LSTM network is introduced and optimized through the GA to address low network efficiency caused by the numerous parameters during training, thereby improving the accuracy of network security situation awareness.

13.21 FNNs FOR ACTIVITY RECOGNITION

Action recognition in video is a significant research challenge for CV researchers. Numerous CV applications exist, including surveillance and monitoring, self-driving cars, sports analytics, and human–robot interaction, among others. Self-driving vehicles greatly depend on this capability, while drones and robots benefit significantly from the advancements being made in the field of computer vision. CNNs play a vital role in various computer vision tasks. Traditional supervised methods require large annotated datasets for training, which are costly and time-consuming to obtain.

Two main research fields focus on the optical flow principle and skeleton joint recognition in action recognition. Optical flow calculation can be demanding in real-time applications. To address this challenge, a novel fractal CNN architecture for action recognition, termed the Teacher Guided Student Network (TGSNet), has been developed. Experiments on the HMDB51 dataset demonstrate the efficiency and effectiveness of the proposed TGSNet, which achieves competitive performance and superior speed compared to advanced methods. Furthermore, this architecture utilizes video clips as input data, incorporating optical flow computation and action recognition within the network. The TGSNet competes well with state-of-the-art simple models using the HMDB51 dataset in experiments. More details can be found in the paper titled "TGSNET: A Fractal Neural Network For Action Recognition."

Several other research publications highlight the growing contributions of FNNs in solving complex problems. One such paper is "FDCNet: Presentation of the Fuzzy CNN and Fractal Feature Extraction for Detection and Classification of Tumors."

13.22 CONCLUSION

Fractals are geometric shapes that can display complex and self-similar patterns found in nature, with clouds and plants being the most prominent examples. A fractal is an infinitely complex shape that is self-similar across different scales, displaying a pattern that repeats endlessly, where every part resembles the whole. Fractals capture the geometric properties of natural elements. Indeed, nature is rich in fractals, such as trees, rivers, coastlines, mountains, and seashells. Despite their complex shapes, fractals can be generated through well-defined mathematical systems, such as IFS. In this chapter, we have discussed FDA and its unique applications. In the latter part of the chapter, we focused exclusively on FNNs and their powerful applications. FNNs have been shown to be ground-breaking and will enhance traditional DNNs such as CNNs and RNNs. FNNs are being positioned as the next generation of DNNs, bringing innovative, disruptive, and transformative use cases across various industry verticals.

REFERENCES

1. Roberto, G. F., et al., Fractal Neural Network: A New Ensemble of Fractal Geometry and Convolutional Neural Networks for the Classification of Histology Images, https://www.sciencedirect.com/science/article/abs/pii/S0957417420308563.
2. Larsson, G., Maire, M., and Shakhnarovich, G., FractalNet: Ultra-Deep Neural Networks without Residuals, https://arxiv.org/abs/1605.07648v4.
3. Narasimhan, R., Fractal Networks for AI, https://www.linkedin.com/pulse/fractal-networks-take-ai-one-step-closer-natural-narasimhan/
4. Sohl-Dickstein, J., The Boundary of Neural Network Trainability is Fractal, https://arxiv.org/pdf/2402.06184v1.
5. Zuev, S., et al., Fractal Neural Networks, https://ieeexplore.ieee.org/document/9698649.

Index

Printed in the United States
by Baker & Taylor Publisher Services